OREGON EPISCOPAL SCHOOL

1869

This book given in honor of

Helen
Van De Water
Trust

In appreciation of their gift to the
Math, Science, and Technology Building

April, 2003

EXPLORING PLANETARY WORLDS

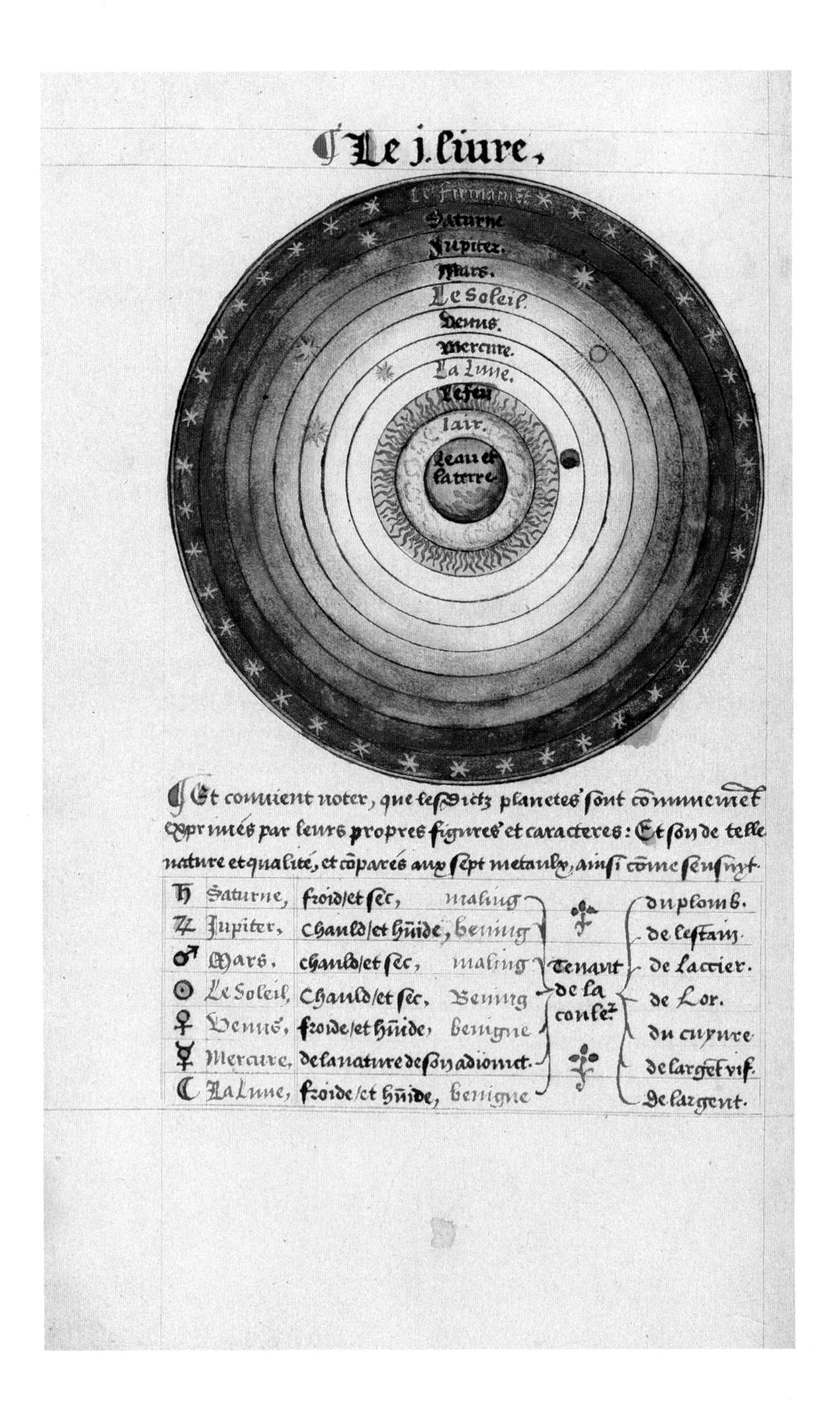

¶ Le j. liure,

¶ Et conuient noter, que lesdictz planetes sont communemẽt exprimés par leurs propres figures et caracteres: Et sont de telle nature et qualité, et cõparés aux sept metaulx, ainsi cõme sensuyt.

♄	Saturne,	froid/et sec,	maling	Tenant de la confe.^z	du plomb.
♃	Jupiter,	Chauld/et huide,	bening		de lestain.
♂	Mars.	chauld/et sec,	maling		de lacier.
☉	Le Soleil,	Chauld/et sec,	Bening		de Lor.
♀	Venus,	froide/et huide,	benigne		du cuyure.
☿	Mercure,	de la nature de son adioinct.			de larget vif.
☾	La Lune,	froide/et huide,	benigne		de largent.

EXPLORING PLANETARY WORLDS

David Morrison

SCIENTIFIC
AMERICAN
LIBRARY
A Division of HPHLP
New York

Library of Congress Cataloging-in-Publication Data

Morrison, David, 1940–
Exploring planetary worlds / David Morrison.
p. cm.
Includes bibliographical references and index.
ISBN 0-7167-5043-0
1. Planets. I. Title.
QB601.M457 1993
523.2 — dc20 92-46641
CIP

ISSN 1040-3213

Frontispiece: The Earth was at the center of the classical universe, as illustrated in this page from a medieval European manuscript.

Printed in the United States of America

Scientific American Library
A division of HPHLP
New York

Distributed by W. H. Freeman and Company
41 Madison Avenue, New York, New York 10010
20 Beaumont Street, Oxford, OX1 2NQ, England

1 2 3 4 5 6 7 8 9 0 KP 9 9 8 7 6 5 4 3

This book is number 45 in a series.

CONTENTS

To two people who have inspired my love of books and motivated me to write them:

Alice Guin, my mother and freshman composition professor
Carl Sagan, my dissertation advisor and friend

PREFACE

Crafting a modern overview of the planetary system has been a challenge and a pleasure. The objective of this book is to reveal the planets as places and to do so from a uniform perspective, taking advantage of the fact that spacecraft have now visited all these worlds except little Pluto. A comparative approach allows us to consider each planet or satellite in context, focusing both on their common origin and on the processes that have molded them. We live in a unique time, at the end of the first reconnaissance of the solar system and poised for the start of more detailed exploration leading in the next century to permanent human presence on the Moon and Mars. As the Russian Konstantine Tsiolkovsky wrote nearly a century ago: "Earth is the cradle of mankind, but we cannot remain in the cradle forever." Other planets beckon, and eventually we will answer that call. In a sense, this volume is a guidebook to that future.

I am grateful to William Kaufmann, who first suggested that I write for the Scientific American Library, and to Jerry Lyons, who invited me to write this book and saw the manuscript through to publication. I have profited greatly from critical reading of the manuscript by Jeff Bell of the University of Hawaii, Dale Cruikshank of NASA Ames Research Center, Bill McKinnon of Washington University, and Janet Morrison of Stanford University. I am also grateful to colleagues at the U.S. Geological Survey, NASA Jet Propulsion Laboratory, and Washington University (St. Louis) for their help in obtaining reprocessed, true-color versions of a number of planetary photographs for use in this book. Amy Johnson edited the manuscript with great skill and provided expert guidance throughout the editorial and production process, while Travis Amos did an excellent job of suggesting and locating photographs for this volume. I sincerely thank all these persons as well as the W. H. Freeman and Company editorial and production team (Georgia Lee Hadler, Megan Higgins, Bill Page, Paul Rohloff, Gail Silver, and Audrey Herbst) for their enthusiastic and highly professional contributions to the creation of this book. It has been a pleasure to work with them.

David Morrison
November 1992

EXPLORING PLANETARY WORLDS

PROLOGUE

WANDERERS

Evolving Perspectives on the Planets

•••

Construction of astronomical observatories has remained an important human activity throughout history, from the stone circles of the ancient Celtic people of Britain (on the facing page: the Stones of Callanish) to the astronomical telescopes of our own time (shown at left: the 10-m Keck Telescope on Mauna Kea, Hawaii).

Within a single lifetime, the skies of Earth have faded. The majestic dome of the heavens, studded with thousands of glittering stars, is a sight denied to most of us today by air and light pollution. Instead of expanding, our visual horizons have shrunk, often to no more than a few city blocks. Occasionally, however, in the desert or on the open ocean, we can still see the glory of the night sky that was the common heritage of our ancestors.

On a clear, moonless night the unaided human eye can see more than 5000 individual stars, in addition to the soft glow of the Milky Way stretching from horizon to horizon. Thousands of years ago, people watched the rising and setting of the stars, identified the brightest by name, and imagined patterns in the sky that related to their myths and gods. In Egypt, the predawn rising of the bright star Sirius signaled the imminent flooding of the Nile and marked the start of a new year. The Greeks used the stars to illustrate the stories of heroes: Perseus, Orion, and Heracles. Tracing the annual cycle of the constellations, the civilizations of the past deduced the apparent path of the Sun against the stars and invented the calendar as a way of regulating agriculture and honoring their gods.

The star patterns are constant. As they cross the sky from hour to hour or shift their time of rising from season to season, the stars remain fixed with respect to each other. Once you have learned to recognize the constellation of Orion, for example, you need not fear that it will change or lose its form. If you travel from the northern to the southern hemisphere you may think that Orion has turned on his head, but it is you, not the stars, who have moved. In contrast,

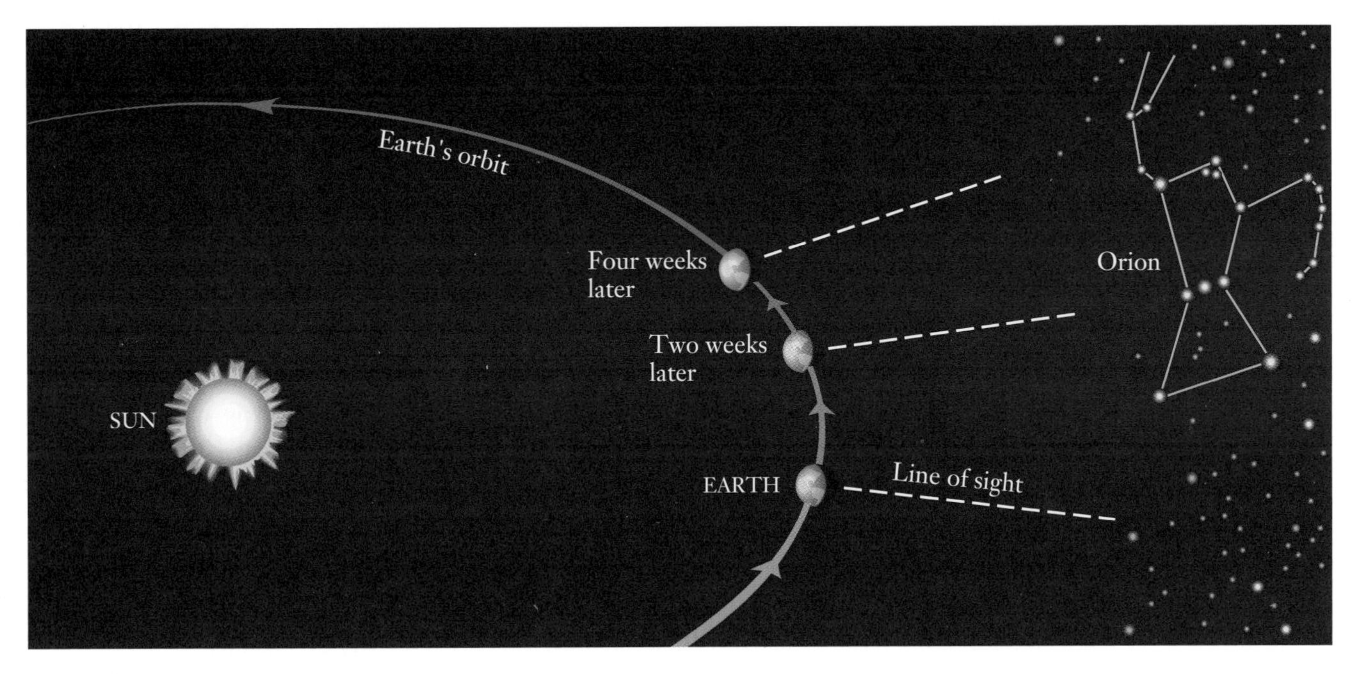

Observed from the moving Earth, the fixed stars rise 4 minutes earlier each night. Over the course of a year, each constellation makes a complete circuit of the sky; Orion (shown here) is best observed in mid-winter, but in summer it lies in the daylight sky and cannot be seen.

however, there are seven objects in the visible heavens that do move against the background of the stars. The ancient Greeks called them planets, or wanderers.

Geocentric Cosmology

The brightest of the ancient wanderers are the Sun and Moon, and their motions were the focus of much early astronomical speculation. Five additional starlike points of light move. These objects, ranked in order of their apparent speed against the background of fixed stars, are called (in our Latin-derived English) Mercury, Venus, Mars, Jupiter, and Saturn. These five planets are, to the naked eye, indistinguishable from stars except for their motion. They are also among the few dozen brightest objects in the night sky, and both Venus and Mars often outshine any star.

All of the ancient civilizations that have bequeathed us written records noted the motions of the planets and made efforts to predict their positions. These peoples assumed that the planets were put in the sky for some purpose and that they influenced events on Earth; in some but not all cases the planets were identified as gods. In Mesopotamia and China, elaborate systems were developed to permit astrologer-priests to interpret planetary positions as signs and omens related to the fortunes of kings and countries. It was not until the Greeks of two thousand years ago, however, that the first astronomers went further and developed detailed theories or models for the motion of the planets.

By the second century A.D., scholars of the Greco-Roman world had developed a consensus out of several centuries of previous observations and theory. Their cosmology—view of the cosmos—was designed to explain the appearance and motion of the sky and, in

In the geocentric cosmology developed by the ancient Greeks, the heavens surrounded the Earth in a series of concentric crystalline spheres, beginning with the Moon (Lunae) and proceeding outward past Jupiter (Iovis) to the sphere of the fixed stars. This Ptolemaic system was widely accepted in Europe and the Middle East.

particular, to provide a way to predict the complicated motions of the planets. This cosmology was consistent with the Platonic philosophy of the time, invoking no gods or miracles but based on the premise that the universe is inherently stable and predictable. This unified cosmological perspective reaches us primarily through the surviving works of the Alexandrian astronomer and geographer Claudius Ptolemaeus, or Ptolemy.

Ptolemy assumed that the cosmos was built of circles and spheres: the spherical Earth in the center, surrounded by concentric spherical shells. The Earth, as the abode of humanity, was fixed and subject to its own natural laws, such as the cycle of birth and death and the attraction of gravity. The celestial realm, in contrast, was made of different stuff and obeyed its own rules, one of which was that all motion was circular and proceeded at constant speed. In order to accommodate the observations of planetary motion, which were obviously not constant, Ptolemy developed a model that combined constant circular motions in ways that would reproduce the observations. His elaborate system of circles accomplished in geometric form approximately the same ends as a modern mathematician who uses algebraic techniques to break down a complex function into an equivalent series of sine and cosine functions. Ptolemy succeeded so well that his model was still being used to predict planetary positions more than a millennium after his death.

In the rest of this chapter we look briefly at how later generations rejected Ptolemy's geocentric cosmology and gradually recognized that, in order to understand planetary motions, we must fundamentally alter our perspective. We have come to understand that the Earth is a planet. The thread that binds this volume together is the effort to understand the planets as places, individual worlds just as varied and fascinating as our Earth.

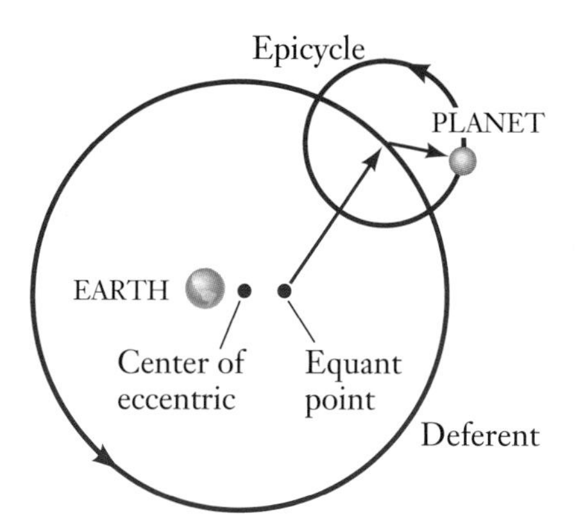

Ancient Greek and Roman astronomers assumed that planetary motions were uniform and circular, yet their observations indicated that the planets did not follow uniform circular paths with the Earth at the center. In the Ptolemaic system, therefore, combinations of circular motions were used to represent planetary motions: for example, placing the planet on a small circle called an epicycle, which rotated about a point on a large circle called the deferent, which in turn was connected to a point called the equant. The system was complicated, but it worked for prediction of planetary positions.

The Earth Moves

The most important intellectual accomplishment of the European Renaissance was the demotion of Earth from the center of the universe. Since the dawn of history people had assumed that the

Earth was the most important part of a universe that had been created for the benefit of humanity. Man was the measure of all things, and he created gods in his image. This ideology was challenged by the concept that the Earth was just one of several planets circling the Sun. With one stroke the Earth became small and the universe large, breaching the barrier between humanity and the cosmos.

This intellectual revolution was initiated by a Polish lay monk, born Mikolaj Kopernik in the Hanseatic city of Torun in 1473. He studied canon law and medicine at Krakow and later at Bologna, Padua, and Rome during a 12-year sojourn in Italy. Leaping beyond contemporary astronomers who sought incremental ways of improving Ptolemy's planetary tables, Copernicus (the Latin form of his name) recognized that planetary motions could be predicted just as well from a heliocentric (Sun-centered) model. Although he could not prove his heliocentric hypothesis, Copernicus made an elegant and persuasive case for his ideas in his opus, *De revolutionibus orbium coelestium (On the Revolutions of the Celestial Spheres)*, published in 1543.

Nicolaus Copernicus (1473–1543) advocated a heliocentric system, in which the planets (including the Earth) orbit the Sun. This concept, that the Earth is one small part of a large and orderly natural universe, had revolutionary implications for both science and philosophy.

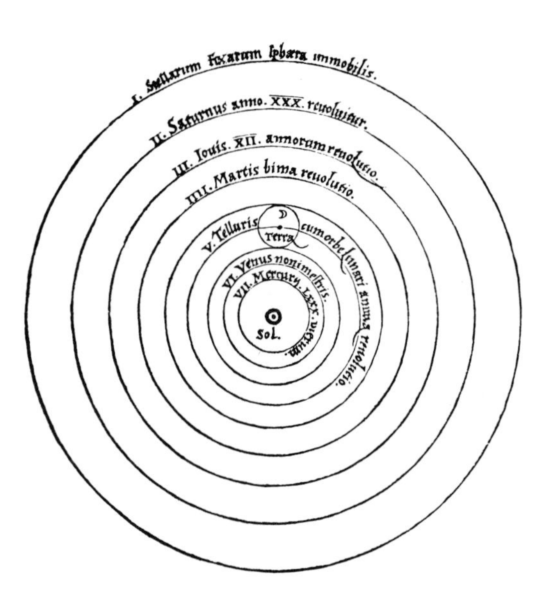

The heliocentric cosmology of Copernicus placed the Sun in the center with the Earth and other planets in orbit about it. In this diagram from De revolutionibus, *the Earth is number V (Telluris), counting from the outermost ring of stars. Since the Moon orbits the Earth, the Copernican system allows for more than one center of motion, thus laying the foundation upon which the concept of universal gravitation would be developed more than a century later.*

Copernicus' work appeared at a moment of great religious and political change, three years before the death of Martin Luther. In Rome, Michelangelo was designing the dome of St. Peter's; in Belgium, Vesalius published his text on human anatomy; on the opposite side of the world, Francis Xavier was discovering the civilization of Japan. It was a propitious time for the introduction of new ideas, and scholars throughout Europe debated the relative merits of the heliocentric and geocentric cosmologies. It was not until the next century, however, that Galileo provided empirical support for the Copernican theory.

Galileo Galilei, born in Pisa in 1564, taught mathematics and astronomy at the universities of Pisa and Padua before becoming scientific advisor to the Grand Duke of Tuscany in 1610. A hero of the scientific Renaissance, Galileo made fundamental contributions to many fields of physics and astronomy, including the first application of the telescope to celestial observation. His discoveries of the phases of Venus and the moons of Jupiter demonstrated the superiority of the heliocentric hypothesis. But perhaps more important than any individual discovery was the synthesis he achieved between theory and experiment. More than any other individual, Galileo

Galileo Galilei (1564–1642) was the father of modern experimental science. He was the first person to use a telescope to observe the heavens, and his discovery of the four moons of Jupiter and the phases of Venus provided strong empirical support for the heliocentric cosmology of Copernicus. His observation that the face of the Moon is rugged and mountainous initiated the study of celestial bodies as worlds like our own Earth.

demonstrated that careful laboratory experiments and observations could be used to establish universal mathematical rules or laws of nature. From our modern perspective, Galileo is the first true scientist.

Galileo's work was suspect in Counter-Reformation Italy, especially because he chose to write in the vernacular about the heliocentric theory of Copernicus and his own discoveries supporting it. In a wave of reaction, the Roman Catholic Church banned Copernicus' *De revolutionibus* and declared it heresy to teach or advocate that the Earth moved. Accused of violating this prohibition, Galileo spent his last years under house arrest. Not until 1992 was Galileo formally exonerated by the Pope.

Johannes Kepler, Galileo's German contemporary, made the critical contributions that placed the ideas of Copernicus on a firm mathematical foundation. A Protestant, Kepler was not subject to the bans of the Roman Catholic Church, but as a young man he was caught in the turmoil of the Thirty Years' War. In 1601, at the age of thirty, he had the good fortune to be appointed court mathematician to the Holy Roman Emperor in Prague. For two

Johannes Kepler (1571–1630) provided the first mathematical description of the orbits of the planets. His laws of planetary motion, determined from a painstaking study of many years of observations (primarily of Mars), allowed the computation of planetary positions with unprecedented accuracy and led to Newton's development of the laws of motion and of gravitation.

decades Kepler carefully analyzed a long series of planetary positions measured by his predecessor, the Danish astronomer Tycho Brahe, in an effort to determine the mathematical laws that govern planetary motion within the Copernican system. Kepler showed that the orbits of the planets are not circles but ellipses, and he discovered the law that relates orbital period to distance from the Sun.

The work of Kepler and Galileo established the validity of the heliocentric cosmology and provided a precise mathematical description of planetary orbits by means of which astronomers could predict the positions of planets with unprecedented accuracy. Equally important, they had abandoned the restrictive assumptions that underlay the Ptolemaic system. For these scientists, the Earth was a planet no different from the other five objects that orbited the Sun. With his telescopes, Galileo had seen mountains and valleys on the Moon: it, too, was a world not unlike our own. Kepler speculated about life on other planets and wrote a science-fiction account of a fantastic trip to the Moon. Although they could not yet reach out

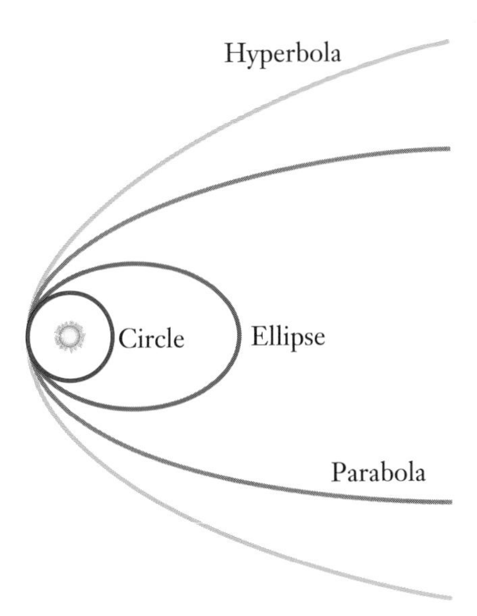

Orbits can follow paths that are circles, ellipses, parabolas, or hyperbolas. Kepler showed that planetary orbits are all ellipses of small eccentricity (i.e., nearly circular).

to touch other worlds, these pioneers allowed humans to begin intellectual exploration of the solar system.

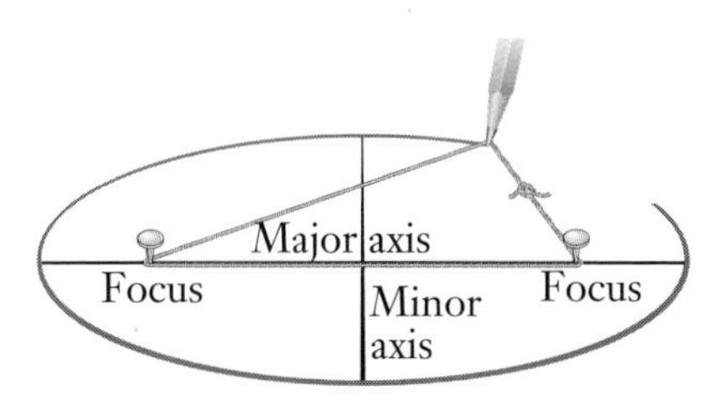

An easy way to draw an ellipse is to loop a string around two tacks. The tacks are at the foci of the ellipse. The closer the two tacks are to each other, the more nearly circular the ellipse is, and the nearer its eccentricity is to zero (the value for a circle). The ellipse illustrated here has an eccentricity of about 0.75.

Laws of Planetary Motion

Kepler's laws of planetary motion provide the tools to understand the structure of the solar system. Kepler derived his laws from observations of Mars and later generalized them to all the planets. Additional generalization incorporates the orbits of planetary satellites, comets, asteroids, and even the paths of interplanetary spacecraft.

Kepler's first law describes the shape of an orbit. From before Ptolemy to after Copernicus, astronomers had assumed that all orbits were perfect circles. The observations of Mars available to Kepler were clearly incompatible with a circular orbit, however, and he therefore tried to match the data with various kinds of elongated ovals. Finally, he hit upon the ellipse. Kepler's first law states that the orbits of planets are ellipses with the Sun at one focus (the other focus is empty).

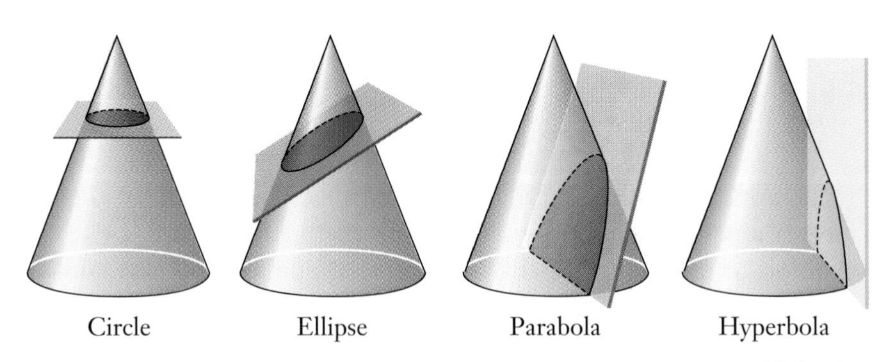

A conic section is the shape that results when a plane and a cone intersect. If the plane is perpendicular to the axis of the cone, we obtain a circle. A sloping plane produces the family of ellipses, with shapes dependent on the angle of the plane. If the plane is parallel to one side of the cone we have a parabola, while even steeper angles result in a family of curves called hyperbolas. Any two bodies moving under their mutual gravitational force will follow paths that are conic sections.

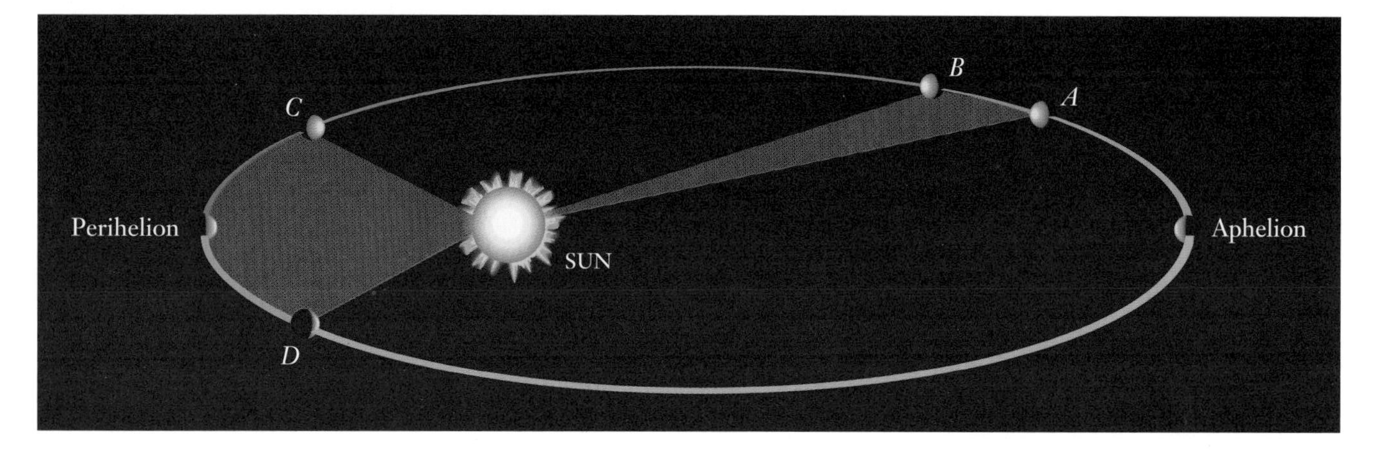

According to Kepler's second law of planetary motion, called the law of areas, for a planet on an elliptical orbit (here highly exaggerated), an imaginary line connecting the planet to the Sun sweeps out equal areas in equal intervals of time. The area of the shaded region between A *and* B *is the same as that of the shaded region between* C *and* D. *Thus the planet accelerates as it nears the Sun (perihelion) and slows when it is far away (near aphelion).*

The circle and the ellipse are members of a family of curves that the mathematicians call conic sections because they are formed by the intersection of a cone and a plane. The shape of the ellipse, measured by its eccentricity — the degree to which the shape is noncircular — is zero for a circle and increases to a maximum value of one for an infinitely elongated ellipse. Later investigators found that certain orbits of spacecraft and comets could be specified by other conic sections, the parabola and the hyperbola, each of which has an eccentricity of one or greater. Today we might restate Kepler's first law as telling us that the paths of two objects moving under their mutual gravitation are always conic sections.

Kepler's second law of planetary motion tells us how fast a planet moves in its orbit. It was apparent to him from observation that Mars moved faster when its orbit brought it closer to the Sun, while it slowed at greater distance. Once he recognized that the orbit was an ellipse, he quickly found that the area swept out by an imaginary line connecting any planet and the Sun is the same in equal intervals of time. When the planet is nearer the Sun it moves faster; even as the line gets shorter, the area swept out is the same.

It required many more years of research before Kepler derived his third law of planetary motion, which relates the average distance of each planet from the Sun to its orbital period. Kepler believed in the underlying harmony of nature, and he considered this law, also

Kepler's harmonic law states that for any planet, the square of the period P *is proportional to the cube of the distance* a *from the Sun. The table shows a test of this law, using modern values for planetary periods (in years) and average distances (in astronomical units).*

EARTH ←1 AU→ SUN

Planet	Average Distance (AU)	Period (Years)	a^3	P^2
Mercury	0.387	0.241	0.058	0.058
Venus	0.723	0.615	0.378	0.378
Earth	1.000	1.000	1.000	1.000
Mars	1.524	1.881	3.537	3.537
Jupiter	5.203	11.862	140.8	140.7
Saturn	9.534	29.456	867.9	867.7

called the harmonic law, to be his greatest achievement. It states that for any planet the square of the period (time elapsed for a complete solar orbit) equals the cube of the distance when the period is measured in years and the distance is expressed in astronomical units (AU). The AU is defined as the distance of the Earth from the Sun, about 93 million miles or 150 million kilometers.

With these laws we can calculate the orbit of each planet about the Sun and thus gain an overview of the solar system. Mercury is the innermost planet, with a period of only 88 days and a distance from the Sun of 0.39 AU. Next is Venus at 0.72 AU, followed by the Earth. Venus and Mercury, with orbits inside the orbit of the Earth and periods of less than a year, are called the inferior planets. The first external (superior) planet is Mars. Taken together, Mercury, Venus, Earth, and Mars are called the inner, or terrestrial, planets; for convenience, we also frequently include the Moon among the terrestrial planets.

A large gap separates the orbit of Mars from that of the next planet outward, Jupiter. Jupiter orbits the Sun in 12 years at a distance of 5.2 AU, nearly a billion kilometers. In the gap between Mars and Jupiter we find many minor planets, or asteroids; all of

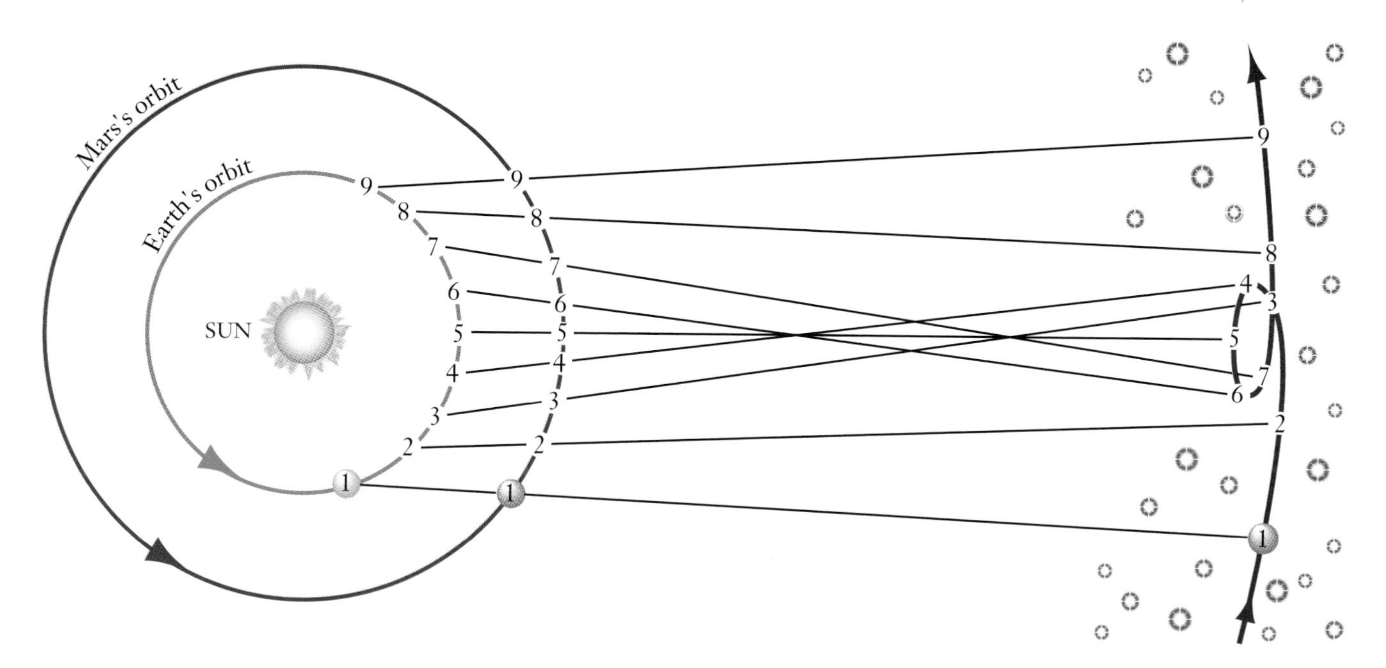

As seen from our own moving Earth, the apparent motion of a planet against the background of stars is complex. Mars, for example, appears to reverse its usual eastward motion each time the faster-orbiting Earth passes it. This retrograde motion is relatively easy to visualize in the heliocentric system.

these combined are less massive than our Moon. Beyond Jupiter lies Saturn, with a period of 30 years (approximately equal to a human generation) and a distance of nearly 10 AU.

Saturn was the most distant planet known to Copernicus or Kepler, but three new planets have been discovered since, all at greater distance. These are Uranus, Neptune, and Pluto. Pluto's distance is 40 AU and its period almost 250 years; it has completed only one-quarter of an orbit since its discovery in 1930. All of the planets circle the Sun in approximately the same plane, like marbles rolling on a table.

Newton and Gravitation

For all their power, Kepler's laws are purely descriptive. They provide a mathematical account of planetary motions, but they do not tell us why the orbits are ellipses or what underlying natural law causes the square of the orbital period to equal the cube of the distance. The next step, a unifying theory of motion both terrestrial and celestial, came with the work of Isaac Newton, perhaps the most brilliant scientist ever to have lived. Many would agree with the couplet by Alexander Pope:

> Nature and Nature's laws lay hid in night:
> God said, Let Newton be! and all was light.

Newton, born in the year after Galileo's death, was a close contemporary of both Tsar Peter the Great in Russia and the German composer J. S. Bach. During most of his career he held the position of Professor of Mathematics at Trinity College, Cambridge, but he was also involved in government service at the time Great Britain was rising to become the first worldwide power. His direct scientific contributions were in mathematics, astronomy, and various branches of physics.

For our purposes Newton's greatest accomplishments were the fundamental laws of motion (still known as "Newton's laws") and the law of gravitation. He recognized that a force was required to alter the speed or direction of any body: to start it, stop it, or bend its path. Thus the ancient Greek concept of perfect circular motion was finally discarded. Newton demonstrated that constant motion in a straight line was the "natural" way, and that an external force was necessary to divert the motion of a planet, for example, from a straight line to a closed curve. This force is gravitation.

Ever since the experiments of Galileo, scientists had understood how the gravitational pull of the Earth influences the motion of falling bodies. Newton's great accomplishment was to establish that this force, felt at the surface of the Earth, actually extends to the Moon and beyond – indeed, all objects exert such a force on each other. He further derived the mathematical form of this force: proportional to the mass of an object and inversely proportional to the square of the distance between objects. With this law Newton showed that the gravitational force of the Earth on the Moon was exactly that required to pull our satellite into a closed path that continually "falls" around the Earth, yet always keeps the same distance.

One of Newton's triumphs was his demonstration that the law of gravitation, as he had formulated it, permitted him to derive Kepler's three laws of planetary motion. He showed that only if the force attracting each planet to the Sun decreases as the square of the

Isaac Newton (1642–1727) was the leading scientist of his century, making fundamental contributions to many disciplines. In addition to formulating his laws of motion and establishing the nature of universal gravitation, Newton invented the reflecting telescope (using a mirror rather than a lens to collect and focus light), which is the standard today for all professional and most amateur observatories.

distance will the orbits be ellipses, the orbital speed follow the second law, and the planets be spaced as described by the third law. Gravitation and the laws of motion are, therefore, the more fundamental concepts, with Kepler's descriptive laws a natural consequence.

Newton's recognition that all objects exert gravitational forces on one other opened the way to a more sophisticated treatment of planetary motion. By calculating the relatively small effects of one planet upon another, scientists could determine their true paths, which deviate slightly from the Keplerian ideal. Precise measurements of planetary positions further allow determination of their individual masses. The mathematical tools created by Newton and his followers are still used today to calculate the trajectories of interplanetary spacecraft. Modern computers are faster and more accurate than anything available to Newton, but the basic principles are the same.

Newton's laws of motion and of gravitation are so general and so powerful that they strongly influenced the way humans thought about the world and their place in it. Many people both within science and outside of it concluded that the basic laws of nature were finally

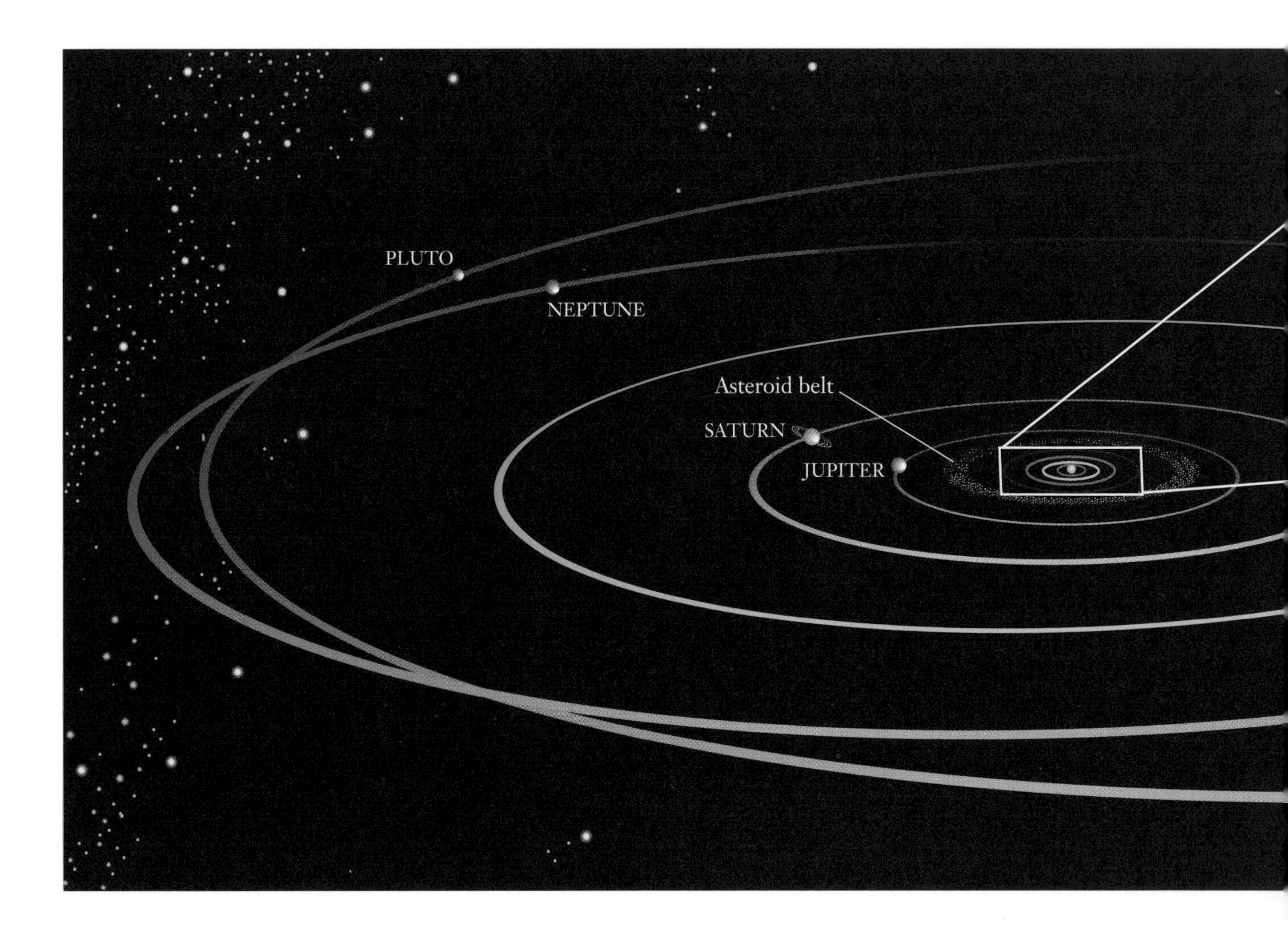

known; it remained only to fill in the details. Some adopted a philosophy of determinism, asserting that every event in the universe can be understood by mechanistic laws applied to the conditions immediately preceding the event. Today few would go so far, but there is no doubt that such thinking has influenced history.

With this background, we are ready to turn to the nature of the planets themselves. Well into the nineteenth century, the work of most astronomers concerned the discovery of new objects and calculations of the orbits of planets, satellites, and the smaller members of the solar system. Only in the past 150 years or so have

Planetary orbits would look like this as viewed at an oblique angle from north of the ecliptic plane, near which these orbits lie. The positions of the planets in this drawing are arbitrary. Note the much larger spacing between orbits for the giant planets, compared with the smaller terrestrial planets near the Sun. While most planetary orbits are nearly circular, the orbit of Pluto is substantially eccentric; in the 1990s, Pluto is actually closer to the Sun than is Neptune.

we shifted emphasis from the motions of these objects to investigation of their physical and chemical nature. In the second half of the twentieth century, our ability to describe and explain the nature of planets as individual worlds has greatly accelerated, culminating in their direct investigation by space probes. Today's planetary scientist asks not only about the *properties* of the planets, but also about the *processes* that have shaped each world to its own special destiny. The following chapters focus on planetary environments and processes — and, ultimately, on what they reveal about the origin and evolution of the solar system.

GIANTS
The Jovian Planets

All of the giant planets of the outer solar system were explored by Voyager 2 (left) between 1979 and 1989. On the facing page we see a Voyager photograph of the crescent Uranus, a view impossible to obtain from Earth.

Before visiting a place for the first time, it is useful to consult a map or atlas. We should approach the planets in the same way. Our intimate views of individual planets will be more meaningful if we begin with a look at the big picture.

The dominant member of the solar system is not a planet, but the Sun itself. Although an ordinary star of moderate mass and luminosity, the Sun is preeminent within the solar system. It has a diameter of about a million miles and is almost a thousand times more massive than all of the system's planets, moons, rings, comets, asteroids, and other debris combined. Possessing such a commanding mass, its gravitation also dominates, and for most purposes we can think of the planets all orbiting an essentially stationary Sun in the center of the system. This is why Kepler's laws work so well, even though he knew nothing of the mutual gravitational attractions of the planets and the smaller members of the system.

Astronomers take the composition of the Sun as a standard against which to compare other members of the solar system. By definition, therefore, the Sun has the so-called cosmic composition that we associate with the birth of the solar system, primarily a mixture of the two lightest elements, hydrogen and helium. Together these make up about 99 percent of the solar mass; within the remaining one percent, the abundant elements include oxygen, carbon, nitrogen, iron, magnesium, and silicon. We now know that the gases hydrogen and helium are primordial, having formed when the universe was born some 15 billion years ago in the big bang. All of the other elements have been manufactured subsequently in the interiors of massive stars and then expelled to enrich (or pollute, depending on your perspective) the interstellar gas and dust from which new generations of stars and planets are born.

The Sun generates energy in its interior by the nuclear fusion of hydrogen into helium. Every second some 600 million tons of hydrogen are converted to helium, with the release of 4×10^{20} megawatts of power. The Sun has been doing this for nearly 5 billion years, and we can expect it to continue for about 5 billion more. This solar energy is the primary source of heat and power for all but the two largest planets.

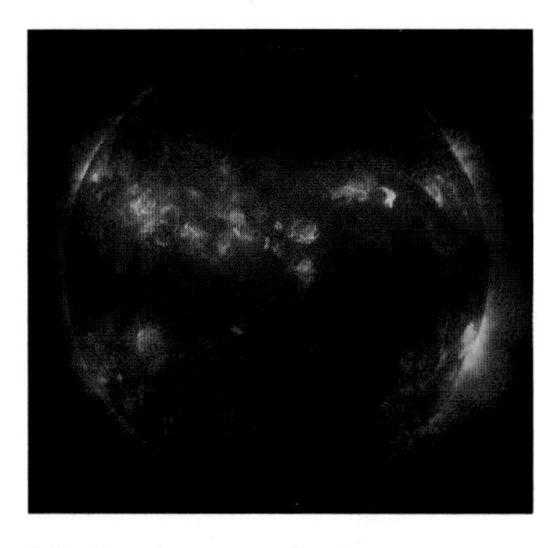

The Sun dominates the planetary system as the center of planetary motion and the source of light and heat for the planets. This orbital photo in X-ray radiation features the million-degree upper atmosphere or corona of the Sun.

Planetary Geography

The dominant members of the solar system after the Sun itself are the giant planets Jupiter, Saturn, Uranus, and Neptune. Jupiter is by far the largest, as befits its association with the king of the Roman gods. Indeed, Jupiter is more massive than the rest of the planetary system put together. Yet it is small in comparison with the Sun. In

Voyager images of Jupiter reveal great complexity in its colorful bands of clouds. This image also shows two of Jupiter's satellites: multicolored Io to the left in front of the Great Red Spot and icy Europa to the right.

round numbers, Jupiter has about one-tenth the diameter of the Sun, therefore about one-thousandth its mass and volume. Saturn is substantially smaller, with only about one-quarter the mass of Jupiter. Although usually classed as giants, the outer two planets, Uranus and Neptune, have masses only about 5 percent as great as Jupiter's.

The four giant planets have rather evenly spaced orbits. Jupiter is on the inside, circling the Sun at a distance of about a billion kilometers. Using the astronomical unit (AU) as our measure of distance, Jupiter is about 5 AU from the Sun. Next is Saturn at about 10 AU, followed by Uranus at 20 AU and Neptune at 30 AU.

After the giant planets, the most massive part of the system is the huge collection of small, icy bodies known as comets. Hovering at the outer fringes of the solar system, the comets number more than a trillion and may total a substantial fraction of the mass of Jupiter itself.

The Earth and other small planets (Venus, Mars, Mercury, Pluto) hardly count in the cosmic order of things. The four terrestrial planets cling close to the Sun at distances of less than 1.5 AU. They

The Planets

Name	Distance from Sun (AU)	Period (Years)	Diameter (km)	Mass (Earth = 1)	Density (Water = 1)
Mercury	0.4	0.24	4,878	0.055	5.4
Venus	0.7	0.61	12,104	0.82	5.3
Earth	1.0	1.00	12,756	1.00	5.5
Mars	1.5	1.88	6,794	0.11	3.9
Jupiter	5.2	11.9	142,800	318	1.3
Saturn	9.5	29.5	120,000	94	0.7
Uranus	19	84	52,400	15	1.2
Neptune	30	165	50,400	17	1.5
Pluto	40	249	2,200	0.002	2.2

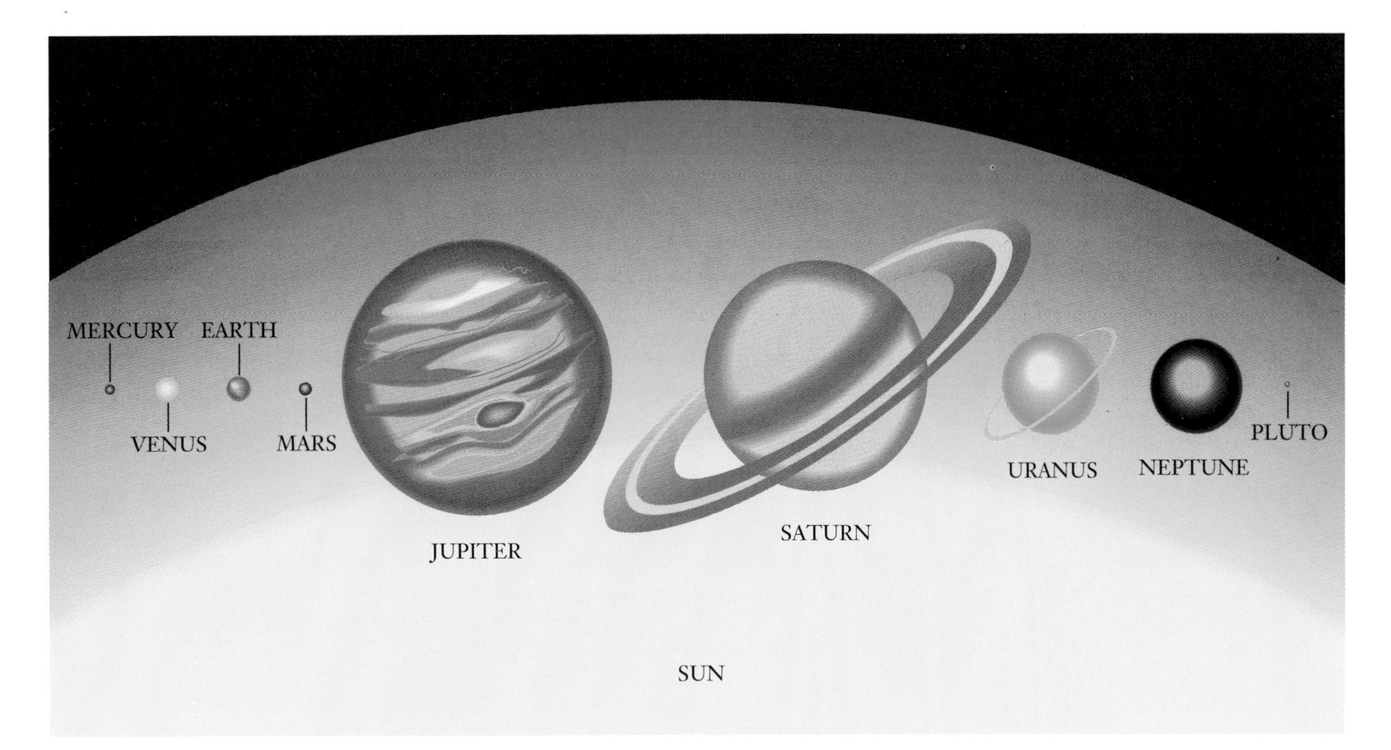

The planets span a wide range in size, but all are small in comparison with the Sun, which has a diameter of about 1 million miles.

are mere cinders of rock and metal, the total mass for all five being only about 1 percent that of Jupiter. To an alien viewing our system from a great distance, these small bodies would be invisible, lost in the glare of the Sun. Our own perspective is different, of course. It is useful, however, to remember that generally the outer planets and their extensive systems of satellites and rings are where most individual objects, as well as most of the mass of the system, are to be found. That is why we start this survey of the planetary system with its dominant members.

To conclude this overview with a mental scale model for the solar system, let us in our imagination reduce all of the dimensions of the solar system by a factor of 1 billion. The Sun is then a sphere nearly 1.5 m in diameter, about the height of an average adult. Jupiter is 5 blocks away, represented by a large grapefruit. The other three giant planets are lemons or oranges at distances of 10 blocks (Saturn), 20 blocks (Uranus), and 30 blocks (Neptune). In the inner solar system, the Earth and Venus are grapes, while Mercury and Mars are more like peas, all orbiting within 1 block of our human-size Sun. In this scale model, a person is reduced to the dimensions of a single atom and automobiles or spacecraft to the size of molecules.

Exploration of the Outer Solar System

Astronomers have learned a great deal about the giant planets from telescopic studies, but much of what we would like to know cannot be discerned from Earth. The answer is to take a close look using robotic spacecraft. Four of these spacecraft have penetrated beyond the asteroids to initiate the exploration of the outer solar system, and a fifth is on the way.

The challenges of probing so far are considerable. Flight times to the outer planets are measured in decades, rather than the few months required to reach Venus or Mars. Spacecraft must be highly reliable and capable of autonomous operation, since light-travel time between Earth and the spacecraft is several hours. If a problem develops near Saturn, for example, the spacecraft computer must deal with it directly; to wait hours for the alarm to reach Earth and instructions to be routed back could spell disaster. These spacecraft also carry their own electrical energy sources, since sunlight at these distances is too weak to supply energy through solar cells. Heaters keep instruments at proper operating temperatures, and large antennas are needed to transmit their precious data to receivers on Earth.

The first spacecraft to the outer solar system were Pioneer 10 and 11, launched in 1972 and 1973 as pathfinders to Jupiter. Their main objectives were to determine whether a spacecraft could navigate through the asteroids without colliding with small particles and to measure the radiation hazards in the magnetic field around Jupiter. Both spacecraft passed through the asteroid belt without incident, but the energetic particles associated with Jupiter nearly wiped out their electronics—providing information necessary to the design of subsequent missions. Pioneer 10 flew past Jupiter in 1974, after which it sped outward toward the limits of the solar system. Pioneer 11 used the gravity of Jupiter during its 1975 encounter to divert it toward Saturn, which it reached in 1979.

The primary scientific missions to the outer solar system were Voyager 1 and 2, launched in 1977. Each carried 11 scientific instruments, including cameras and spectrometers as well as devices to measure planetary magnetic fields. Voyager 1 reached Jupiter in 1979 and used the gravity of that planet to boost it on to Saturn in 1980. The second Voyager, arriving at Jupiter four months later, followed a different path to accomplish a grand tour of the outer planets: Saturn in 1981, Uranus in 1986, and Neptune in 1989. Many of the stunning color photographs in this book were obtained with the Voyager cameras during these flybys.

Voyager 2 followed a trajectory made possible by a rough alignment of the four giant planets on the same side of the Sun. About once every 175 years, these planets are in a position where a single spacecraft can visit them all, using gravity-assisted flybys to adjust its course for the next encounter. Because of this alignment,

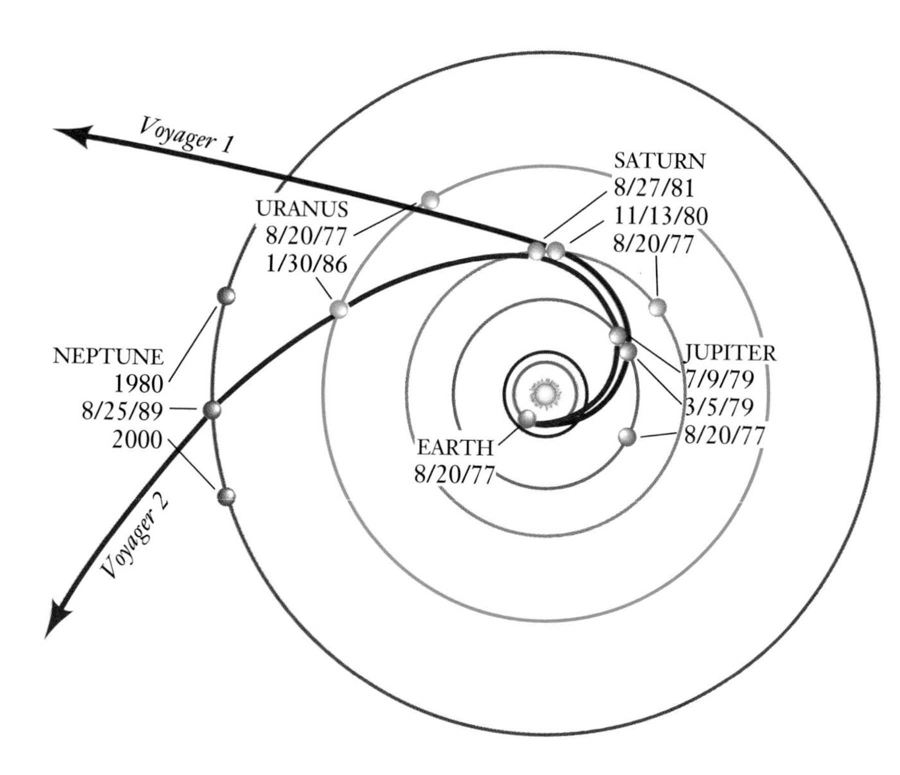

The flights of the two Voyager spacecraft through the solar system represent triumphs of celestial navigation. Launched in 1977, both spacecraft used the gravity of Jupiter to accelerate them on toward Saturn. The Voyager 2 aimpoint at Saturn provided another gravity assist to Uranus, and from there on to Neptune. Voyager 1 was targeted differently at Saturn, sacrificing its grand tour for close views of the satellite Titan. Now both spacecraft have achieved escape velocity with respect to the Sun, and both are leaving the solar system, never to return.

Missions to the Outer Solar System

Encounter Date	Planet	Spacecraft
1973	Jupiter	Pioneer 10
1974	Jupiter	Pioneer 11
1979	Jupiter	Voyager 1
1979	Jupiter	Voyager 2
1979	Saturn	Pioneer 11
1980	Saturn	Voyager 1
1981	Saturn	Voyager 2
1986	Uranus	Voyager 2
1989	Neptune	Voyager 2
1995	Jupiter	Galileo
2004	Saturn	Cassini

every planet in the outer solar system except Pluto has been visited by spacecraft; otherwise, it would probably have been well into the next century before this basic reconnaissance of the planetary system could be accomplished.

Astronomers agree that the next steps in the exploration of the outer solar system involve extended study of Jupiter, Saturn, and their satellites. The Galileo mission to Jupiter, launched in 1989, will arrive in 1995 and deploy an entry probe for direct studies of the atmosphere; it will then begin a three-year orbital tour including repeated close flybys of the four large jovian satellites. The similar Cassini mission to Saturn is under development as a cooperative venture between NASA and the European Space Agency. If all goes well, Cassini will be launched in 1998 and arrive at Saturn in 2004. Meanwhile, we continue to study these objects with terrestrial telescopes and to analyze the wealth of data sent back by the two Voyagers.

In 1995, the Galileo spacecraft will release a probe into the atmosphere of Jupiter. This artist's impression shows the instrumented capsule separating from the heat shield that has protected it during its fiery entry into the jovian atmosphere.

Planetary Composition

One of the greatest achievements of twentieth-century astronomy is the ability to measure the composition of celestial objects. Using spectrometers to sort the light from distant sources by wavelength, we have learned to recognize the unique spectral signatures of various elements and compounds. Thus we can decipher the composition of any luminous source without ever touching it or subjecting its material to laboratory analysis.

Spectroscopy is the chief tool of the modern astronomer, and more than half the observing time of large telescopes is devoted to spectral studies. Spectrometers are also among the basic instruments carried on planetary spacecraft like Voyager and Galileo. We can use spectroscopic techniques to probe the compositions of nearby planets as well as distant stars and galaxies.

These techniques are particularly useful for the analysis of the gas in planetary atmospheres. Gases tend to imprint very specific spectral

Most of the information that astronomers derive from the light of distant objects is obtained by breaking the light into its spectral components. The electromagnetic spectrum is the sum of all of these components, each with different wavelength and energy. The illustration shows the electromagnetic spectrum stretching from the highest energy gamma rays down to very low energy radio waves with wavelengths of tens of kilometers. The Earth's atmosphere is opaque to X rays and gamma rays and to a large part of the infrared spectrum. To observe in these spectral regions, we must place our telescopes above the atmosphere.

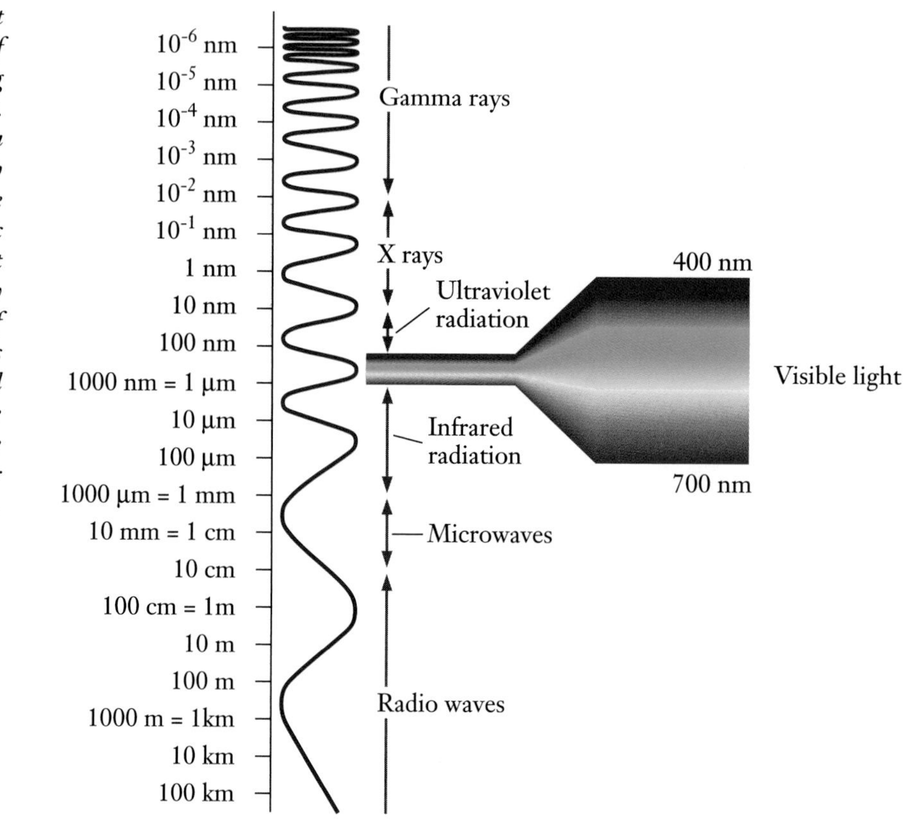

Numerous absorption lines are seen in this photograph of the Sun's spectrum. Each dark absorption line is the spectral signature of an element in the Sun's upper atmosphere. The spectrum is so long that it had to be cut into segments to fit on this page.

patterns upon any light passing through them, and these patterns tell us the detailed abundances of various compounds. There are complications, of course. The sunlight that strikes a planet already carries the spectral signatures of the gases in the Sun, to which the information about the planetary atmosphere is added. Before it reaches our telescopes, the same beam of light passes through the atmosphere of the Earth, where it is further modified. Most of these problems have been overcome, however, so that we can distinguish the spectral imprint of a planetary atmosphere from that of the Sun or Earth. In a few critical cases – for example, to measure the helium

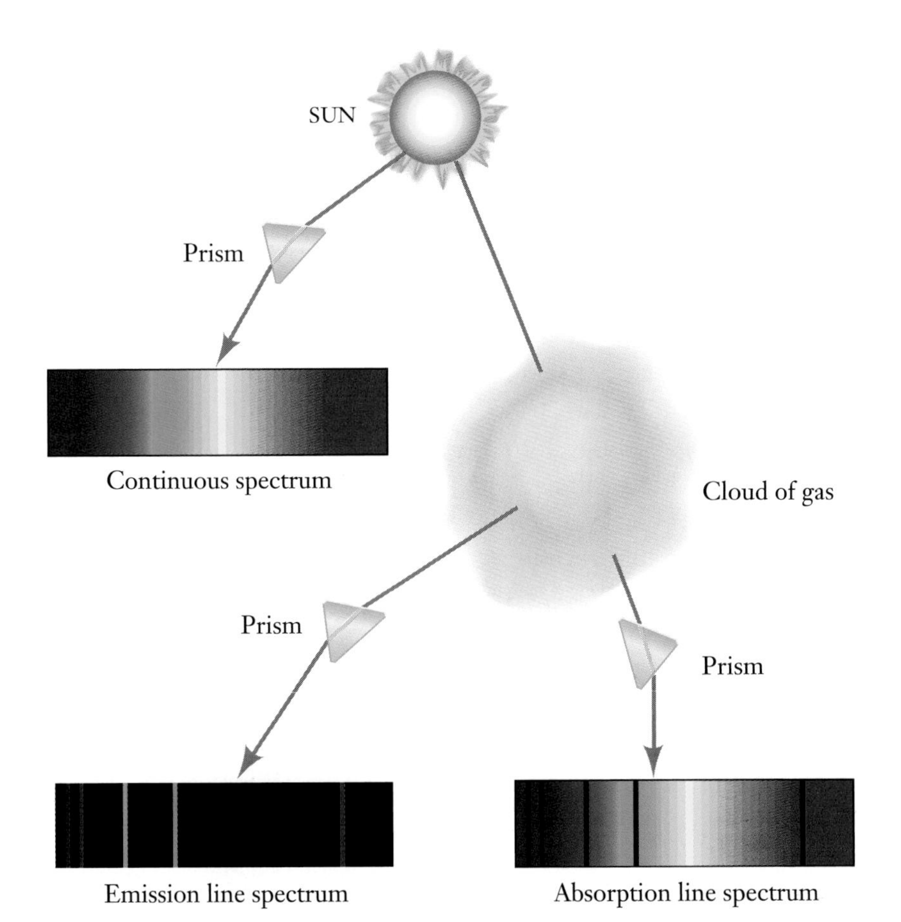

The analysis of spectra allows the astronomer to determine the composition and temperature of distant objects. The basic laws of spectral analysis were discovered in the late nineteenth century. A glowing solid or dense gas emits a continuous spectrum, with all colors present. If this light passes through a dilute gas (like a planetary atmosphere), part of the light is absorbed at discrete wavelengths (colors) that are diagnostic of the composition of the absorbing gas. A dilute gas also can emit light at exactly the same discrete wavelengths, to generate an emission spectrum. Either the absorption or emission spectra can be used to determine the properties of the gas.

abundances in the outer planets — we have our best results from spectral measurements made by spacecraft outside the atmosphere of the Earth. Our knowledge of the atmospheres of Mars and Venus was further enhanced by direct measurements made from spacecraft that descended through their atmospheres or landed on their surfaces.

Determining the composition of the solid or liquid parts of a planet is more difficult. Spectra of the light reflected from a solid surface provide some information (iron oxides are orange, for example, and water is blue), but rarely can the complex compounds and minerals be clearly identified and their relative abundance measured. There is quite a bit of educated guesswork in determining the composition of planetary surfaces from remote observations. Even more problematic is the question of the composition of planetary interiors. We can only measure the parts of the atmosphere or surface that we can see, and the detailed nature of the invisible bulk of a planet is sometimes rather speculative.

One of the simplest ways to assess the bulk properties of a planet is to determine its density — that is, the ratio of mass to volume. As we all know, different materials have different densities. Just compare a lead brick with a ceramic brick with a styrofoam brick; they may all have exactly the same size and volume, but their densities are dramatically different. The same reasoning is applied to a planet by comparing its density with rock, metal, or ice. For example, from its observed density of about 3 g/cm^3, similar to terrestrial rock, we can conclude with some confidence that the Moon is made primarily of rocky material and lacks a dense metallic core.

We are reasonably certain that all of the planets and most of the smaller members of the solar system formed at about the same time and place as the Sun, so that everything began with the same stock of raw materials. Yet it is evident that the planets did not all emerge from the formation process with the original solar (cosmic) composition. The Earth, for example, is composed mostly of the elements silicon, iron, oxygen, and aluminum — all minor constituents of the Sun. Some process, probably associated with planetary formation, selected some of the raw materials but rejected others. Planetary scientists call this process fractionation. Not only the composition but also many other properties of each individual planet, satellite, or asteroid can be traced back to the chemical fractionation that it experienced at birth or subsequently in its evolutionary history.

The dominant members of the planetary system—Jupiter and the other three giant planets—are the members that most closely approximate the cosmic composition. Jupiter, in particular, seems to have very nearly the same composition as the Sun, made up primarily of the light gases hydrogen and helium. Saturn has almost the same composition, but Uranus and Neptune are depleted of these two gases relative to heavier material.

The Sun and planets formed from the solar nebula, a spinning disk-shaped cloud of gas and dust. Initially, the disk probably had the same composition as the Sun. The planets formed from solid material in the disk, as we will discuss in greater detail later. Every planet has a core of rocky and metallic compounds, but only Jupiter succeeded completely in attracting and holding to this heavy core a large

Cosmic Proportions of the Ten Most Abundant Elements

Atomic Number	Element	Symbol	Number of Atoms per Million Hydrogen
1	hydrogen	H	1,000,000
2	helium	He	80,000
6	carbon	C	450
7	nitrogen	N	92
8	oxygen	O	740
10	neon	Ne	130
12	magnesium	Mg	40
14	silicon	Si	37
16	sulfur	S	19
26	iron	Fe	32

When we look at their relative sizes and internal structures, we see that the planets Uranus and Neptune really are quite different from the larger Jupiter and Saturn. However, each of the four giant planets has a similar core with a mass about 10 times the Earth's mass. The diffrences among these planets derive primarily from the amounts of hydrogen and helium that they were able to attract and hold as they formed.

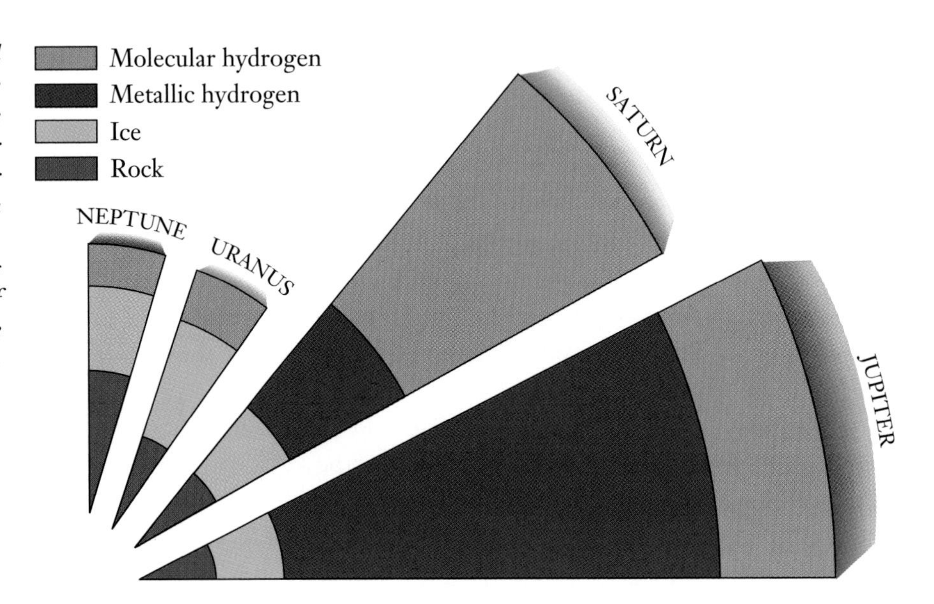

gaseous envelope of hydrogen and helium. The other three giants managed to hold some hydrogen and helium, but less than Jupiter; this explains in part why they ended up less massive than Jupiter. The inner planets missed out on the light gases completely, but that is a story we will tell in subsequent chapters.

Where Hydrogen Rules

Hydrogen, the most abundant element in the outer solar system, is also one of the most chemically reactive gases we know. In the presence of hydrogen, most other elements form compounds that include hydrogen, a condition the chemist refers to as reducing. Consequently, we anticipate that the chemistry of the outer solar system should be dominated by hydrogen-rich (reduced) compounds.

The situation is different in the inner solar system, where little hydrogen was retained by the terrestrial planets. Most of the elements that can combine with hydrogen to produce reduced compounds can also react with oxygen in the absence of hydrogen to make oxidized compounds. The chemistry of the solar system is thus divided neatly into two regimes: hydrogen-rich (reducing) in the outer solar system and oxygen-rich (oxidizing) in the inner solar system.

Some examples help to illustrate the difference between reducing and oxidizing conditions. Two of the most common (and important) elements in the universe are carbon (C) and nitrogen (N). In the presence of hydrogen (H), they form reduced (hydrogen-rich) compounds such as ammonia (NH_3), methane (CH_4), various hydrocarbons (ethylene, C_2H_4; ethane, C_2H_6; propane, C_3H_8); or cyanide (HCN). In an oxidizing (oxygen-rich) environment, however, we get a different set of compounds of carbon and nitrogen, such as carbon dioxide (CO_2), carbon monoxide (CO), or nitric oxide (NO). Oxidizing conditions also lead to the formation of the common minerals – most composed of oxygen in combination with silicon, iron, magnesium, and aluminum – that make up the rock on Earth and the other terrestrial planets.

Whenever both oxygen and hydrogen are present, they will combine to form H_2O, water. Generally, the formation of water takes precedence over other chemicals. Therefore in a hydrogen-rich environment nearly every oxygen atom will end up in a water molecule, and only the excess hydrogen is available to form reduced compounds of carbon, nitrogen, or other elements. If oxygen predominates, most of the hydrogen atoms will combine with oxygen to form water, with the excess oxygen available to make other oxidized compounds. Either way we tend to get a lot of H_2O, unless hydrogen is simply not present in significant quantities.

With this orientation, we can begin to anticipate the situation on Jupiter and other bodies that formed where hydrogen was plentiful. After hydrogen and helium gases, we expect that one of the most common chemicals will be water. In a gas mixture of cosmic or solar composition, there are about 1400 hydrogen atoms for every atom of oxygen. Each oxygen atom combines with two hydrogen atoms to form H_2O, thereby using up all the available oxygen but only one in 700 of the hydrogen atoms. Another abundant compound will be methane. With one carbon atom for each 2000 hydrogen atoms, we can use up all the carbon to form methane (CH_4) and still have plenty of hydrogen left over. Continuing through the periodic table of elements, we can make many reduced compounds; but always there is more unattached hydrogen gas, assuming that we start with the cosmic proportions of the elements.

What does this suggest about the nature of Jupiter? If it has cosmic composition, it will mostly be hydrogen and helium gases. The most abundant solids will be (if the temperature is low enough) common ices like water and methane. Mixed into the atmospheres we can expect a wide variety of reduced compounds, including hydrocarbons and other carbon-based chemicals that we on Earth call organic. (Of course, the organic compounds on Jupiter are not the product of biological activity, but simply the natural result of reducing chemistry.) The same should apply to Uranus and Neptune, but with less hydrogen and helium gases and relatively more solids such as water and methane ice. We shall now examine the evidence to see if these ideas are confirmed by observations.

Jupiter: Biggest Giant

Jupiter is truly the king of the planets. It is 318 times as massive as the Earth and has a diameter of 142,800 km, about 11 times that of the Earth. As seen with a telescope or through the camera of a spacecraft, its colorful and dynamic appearance tells us that we are looking at an object with an extensive atmosphere. Indeed, when we look at any of the giant planets, we see only their atmospheres; if any solid surface exists, it is hidden by opaque clouds.

Motions in Jupiter's cloud patterns allow us to determine the rotation rate of the atmosphere at the cloud level, although this apparent rotation of the atmosphere may not be exactly the same as the spin of the underlying planet. More fundamental is the rotation of Jupiter's core, which can be determined by measuring periodic variations in the magnetic field. Since the magnetic field originates deep inside the planet, it shares the rotation of the interior. This magnetic rotation period of almost 10 hours gives Jupiter the shortest day of any planet. Furthermore, Jupiter has no seasons to speak of because its axis of rotation is tilted by only 3 degrees.

Astronomers are confident that the interior of Jupiter is composed primarily of hydrogen and helium. These gases have been measured only in the atmosphere, but calculations first carried out

Our first detailed views of the atmosphere of Jupiter were obtained by Voyager in 1979. The banded structure of the clouds reflects strong east-west winds in the atmosphere. Two galilean satellites (Io and Europa) can also be seen.

50 years ago show that they are the only possible materials out of which a planet with the observed mass and density of Jupiter could be constructed. Both laboratory measurements and atomic theory tell us how these gases behave at the high temperatures and tremendous pressures in the interior of Jupiter. We find that at depths of only a few thousand kilometers below Jupiter's visible clouds, pressures become so great that hydrogen changes from a gaseous state to a liquid; still deeper, this liquid hydrogen can conduct heat and electricity like a metal. The greater part of Jupiter's interior is liquid metallic hydrogen. Frequently the planet is referred to as a gas giant, but it would be more appropriate to call it a liquid giant—an immense drop of compressed hydrogen and helium.

It is interesting to note that Jupiter has very nearly the maximum possible size for a body of "cold" hydrogen, that is, one that is not generating energy as does a star. Less massive bodies than Jupiter

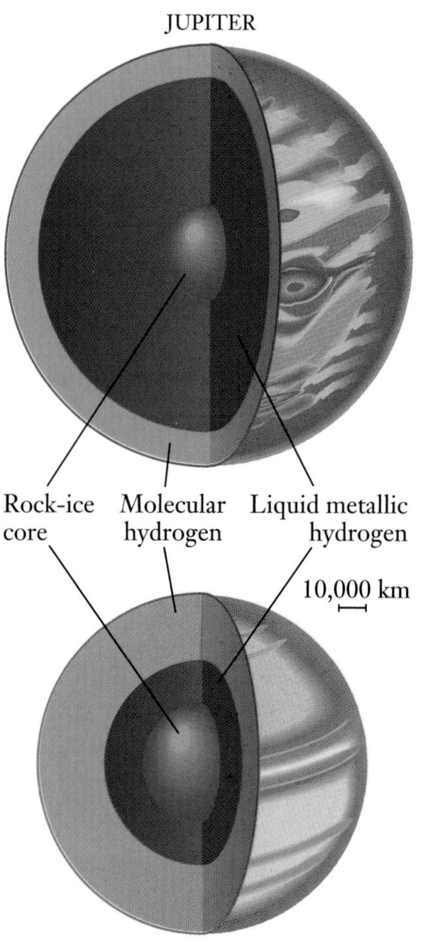

Jupiter and Saturn have similar internal structures. Both are primarily liquid. Each has a hot core of metal, rock, and ice with a total mass about 10 times the mass of the Earth, surrounded by a massive liquid mantle of hydrogen and helium. Within Jupiter, most of this hydrogen mantle has the electrical properties of a metal, while Saturn, with its lower mass and internal pressure, has a much smaller region of metallic hydrogen.

would occupy a smaller volume (like Saturn). More massive bodies, by virtue of their greater gravitation, would be compressed to a smaller volume than Jupiter's. Such an object, its mass larger than Jupiter's but not large enough to maintain nuclear reactions, is called a brown dwarf or infrared dwarf.

Because of their large sizes, each of the giant planets was heated during its formation by the collapse of surrounding nebular gas onto its core. Jupiter, being the most massive, was by far the hottest. In addition, it is possible for giant, largely gaseous planets to generate heat after formation by slow contraction. The effect of these internal energy sources is to raise interior temperatures to values that are higher than would be possible by solar heating alone.

Jupiter has the largest internal energy source, amounting to 400 billion megawatts. It is glowing with the equivalent of 4 million billion hundred-watt light bulbs—about the same as the total solar energy absorbed by Jupiter. The atmosphere of Jupiter, then, combines properties of a normal planetary atmosphere, which obtains most of its energy from the Sun, with those of a stellar atmosphere, heated from below by an internal energy source. Most of the internal energy of Jupiter is primordial heat, left over from the formation of the planet 4.5 billion years ago.

Although it has a substantial internal energy source, Jupiter is in no danger of turning into a luminous star. Stars generate energy by the fusion of hydrogen to helium, a process that requires core temperatures of millions of degrees Kelvin. Such temperatures, in turn, require a total mass of at least 10^{26} tons. Jupiter falls short of this minimum mass to initiate self-sustaining nuclear reactions by about a factor of 70. Jupiter is not a failed star, simply a large planet.

If planets formed the same way as stars, we would expect to find many objects in the Galaxy more massive than Jupiter but smaller than the minimum stellar mass. The fact that such brown dwarfs have not been found in the solar neighborhood suggests that the planet-forming process is quite different from that which generates stars of a wide variety of masses. Although there are many double stars—two stars orbiting each other—they probably were not formed the way that Jupiter and the other planets condensed out of the disk-shaped solar nebula.

Atmosphere and Clouds

The parts of the jovian planets that we can observe directly are their atmospheres. Early in this century, spectroscopic observers identified the presence of ammonia and methane. At first it was thought that these gases might be the primary atmospheric constituents, but now we know that hydrogen and helium are the dominant gases. Not until the Voyager spacecraft measured the far-infrared spectra of Jupiter was a reliable abundance for helium determined.

The clouds of Jupiter are among the most spectacular sights in the solar system, much beloved by makers of science-fiction films. They range in color from white to orange to blue to brown, swirling and twisting in a constantly changing kaleidoscope of patterns. At the temperature and pressure of the upper jovian atmosphere, methane remains a gas but ammonia can condense, just as water vapor condenses in the Earth's atmosphere, to produce clouds. The primary clouds that we seen when we look at Jupiter are composed of crystals of frozen ammonia.

The temperature near the jovian cloud tops is about −130 C, but rises as we penetrate the atmosphere. Through breaks in the ammonia clouds, we can see other layers of cloud that exist in these deeper regions that will be sampled directly by the Galileo probe in 1995.

The four giant planets all have similar atmospheres, composed primarily of hydrogen and helium. In these four diagrams, the altitudes are referred to an arbitrary level where the pressure is 0.1 bar. In each case the temperature is near its minimum at the top of the upper deck of clouds. On Jupiter and Saturn this visible cloud deck is composed of ammonia crystals, while the visible clouds on Uranus and Neptune are made of frozen methane. The deep atmospheres of Uranus and Neptune are left blank to illustrate how little we know about these regions.

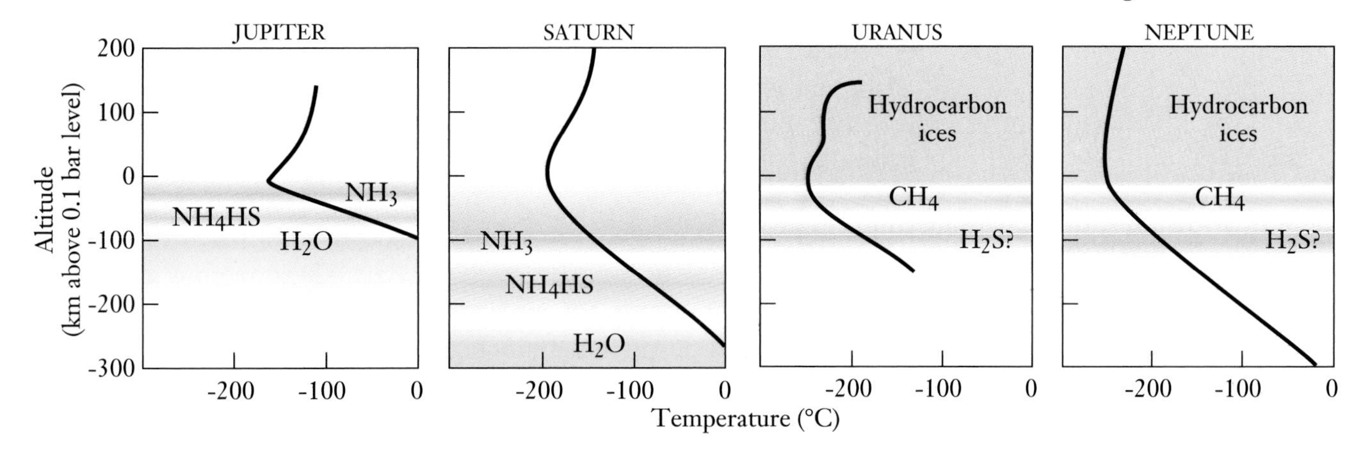

Below the ammonia clouds this probe should pass through a clear region before entering another thick deck of condensation clouds, composed of ammonium hydrosulfide; they probably also contain some sulfur particles, as we judge from their dark yellow or brown color. The Galileo probe should pass next into a region of frozen water clouds, and below that into clouds of liquid water droplets perhaps similar to the clouds on Earth. This region corresponds almost to a shirtsleeve environment on Jupiter, in which hypothetical astronauts could exist quite comfortably if they carried scuba gear for breathing. But with no solid surface to stop it, the probe will continue to descend to dark regions of higher and higher pressure and temperature. No matter how strongly it was built, eventually the probe will be crushed and swallowed in the black depths, where great pressures finally transform the atmospheric hydrogen into a hot, dense liquid.

Although we think we know the composition of the main clouds on Jupiter, we still do not understand their colors. Ammonia condensation clouds should be white, like water clouds on Earth, yet we see beautiful and complex patterns of red, orange, and brown. Some additional chemical or chemicals must be present to lend the clouds such colors, but we do not know what they are. Various organic compounds have been suggested, as well as sulfur and red phosphorus. But there are no firm identifications, nor any immediate prospect of solving this mystery.

Observations of the changing cloud patterns in the atmosphere of Jupiter permit us to measure wind speeds and track atmospheric circulation. The main features of the visible clouds are alternating dark and light bands that stretch around the planet parallel to the equator. Generally, the light zones on Jupiter are regions of upwelling gas, capped by white ammonia cirrus clouds; they apparently represent the tops of upward-moving convection currents. The darker belts are regions where the cooler atmosphere moves downward, completing the convection cycle. These regions are dark because there are fewer ammonia clouds and it is possible to see deeper, perhaps down to the ammonium hydrosulfide clouds.

Superimposed on the regular atmospheric circulation patterns described above are many local disturbances — weather systems or storms, to borrow terrestrial terminology. The largest and most

The Great Red Spot of Jupiter, 30,000 km across, is the largest atmospheric feature on the planet. The jovian jet streams passing near the Red Spot break up into complicated patterns of atmospheric turbulence. The White Oval below the Red Spot is an atmospheric feature that formed in 1940.

famous storm on Jupiter is the Great Red Spot, a reddish oval in the southern hemisphere that is almost 30,000 km long—big enough to contain two Earths side by side. First seen 300 years ago, the Red Spot is clearly much longer-lived than storms in our own atmosphere. It also differs from terrestrial storms in being a high-pressure region. Three similar but smaller disturbances, called the White Ovals, formed on Jupiter in about 1940.

We don't know exactly what caused the Great Red Spot or the White Ovals, but it is possible to understand how they last so long once they form. On Earth a large oceanic hurricane or typhoon typically has a lifetime of a few weeks, even less when it moves over the continents and encounters friction with the land. On Jupiter there is no solid surface to slow down an atmospheric disturbance, and furthermore the sheer size of these features lends them stability. It is possible to calculate that on a planet with no solid surface, the lifetime of anything as large as the Red Spot should be measured in centuries, while lifetimes for the White Ovals should be measured in decades—consistent with what we have observed.

Lord of the Rings

The most spectacular attribute of the second planetary giant—Saturn—is its beautiful system of rings, which sets it apart from all the other planets. But for now we will try to forget the rings, which we defer to a later chapter, in order to focus on the planet itself and to compare it to its larger cousin, Jupiter.

Saturn has about one-fourth the mass of Jupiter, but it is nearly as large, with a diameter of 120,000 km. The planet rotates in about 11 hours, a little longer than the jovian day, and its axis of rotation is tilted by 27 degrees. This tilt, similar to the tilt of the Earth's axis, induces seasons on Saturn. The planet's composition seems to be very similar to that of Jupiter. Because Saturn is farther from the Sun, its temperatures are somewhat colder, and its atmosphere is more extended than Jupiter's as the result of lower surface gravity. However, the same general structure and the same condensation clouds are present.

Voyager obtained this view of Saturn as it approached the planet in 1981. The clouds of Saturn are less colorful than those of Jupiter, but the structure and dynamics of the atmosphere are similar. Three of the satellites of Saturn appear in the bottom of the image.

Yet the appearances of the two planets are different. Saturn lacks the dynamic and colorful cloud patterns of Jupiter; instead it normally displays a bland face with a faint butterscotch hue. The problem is that our view is partly blocked by a hazy saturnian upper atmosphere that makes it difficult to see the underlying cloud patterns. Where the Voyager spacecraft were able to photograph these patterns, they found many similarities to Jupiter. Although there were no storm systems to compare with the Great Red Spot, the wind speed in Saturn's jet streams is actually higher than at Jupiter.

Saturn has an internal energy source about half as large as that of Jupiter, which means (since its mass is only about one-quarter as great) that it is producing twice as much energy per ton of material as does Jupiter. Being smaller, Saturn is expected to retain less primordial heat than Jupiter, so there must be another source for most of this power. This source is believed to be the separation of helium from hydrogen in the interior, where the heavier helium forms drops that sink to release gravitational energy. This precipitation of helium is possible in Saturn because it is cooler than Jupiter; at the temperatures in Jupiter's interior, hydrogen and helium remain well mixed. This theory was apparently confirmed when Voyager measured a lower abundance of helium in the atmosphere of Saturn than had been found for Jupiter. This missing helium has already "rained" into the interior.

Beyond Jupiter and Saturn we find Uranus and Neptune, the little giants – roughly intermediate in size between the two real giants and the terrestrial planets like Earth.

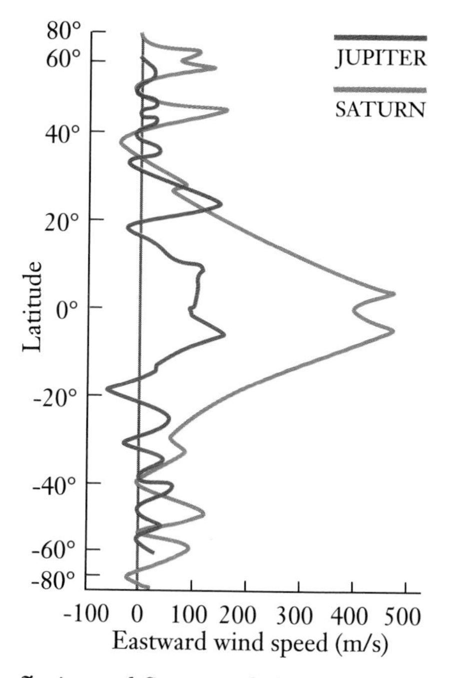

Jupiter and Saturn each display a stable pattern of eastward blowing winds. This graph shows how the wind speed near the cloud tops varies with latitude for both planets.

A World Turned Upside Down

Uranus was the first planet to be discovered telescopically. It was found in 1787 by the German-born English musician and amateur astronomer William Herschel, who was making a systematic telescopic survey of the sky. Herschel saw an object through his telescope that appeared larger than a stellar image; thinking it to be a comet, he followed its motion for several weeks. When an orbit was computed, however, the object could be identified as a planet,

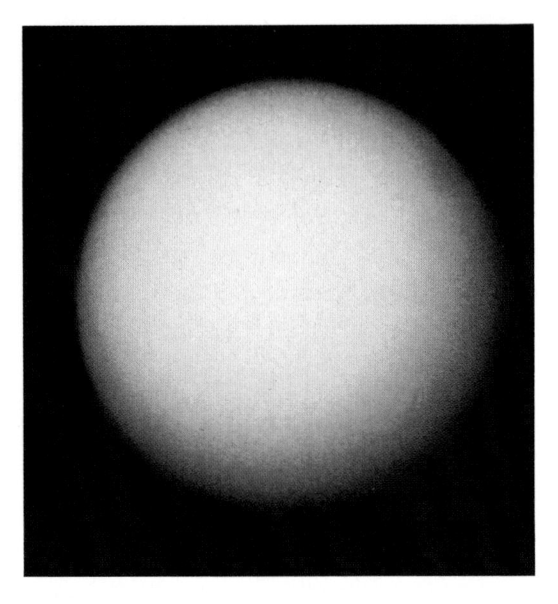

The disk of Uranus is virtually featureless, as shown in this Voyager image taken in 1986 when the rotation pole of Uranus was tipped toward the Sun. Absence of an internal heat source contributes to this bland exterior.

moving in a nearly circular orbit beyond Saturn. Herschel proposed naming his discovery Georgium Sidus (George's Star) after the reigning monarch of England. This astute political move contributed to Herschel's being granted a lifetime research position by King George III, but the international astronomical community insisted that the new planet be named Uranus.

Uranus has a much smaller mass than Jupiter but also a higher density, demonstrating that it must be composed of heavier materials than the hydrogen and helium that dominate on Jupiter and Saturn. Careful examination reveals that the rocky and icy core of Uranus is nearly as large as that of Jupiter or Saturn; what is lacking is a deep envelope of hydrogen and helium, elements that make up less than a quarter the mass of either Uranus or Neptune.

One of the most surprising aspects of Uranus is its rotation. Unlike Jupiter, Saturn or, indeed, any other planet except little Pluto, Uranus has an axis of rotation tilted on its side. This unusual orientation results in very strange seasons, with each pole alternately tipped toward the Sun for 40 years at a time. The planet's rotation period is about 17 hours, longer than for Jupiter or Saturn but still shorter than the day on Earth.

Unlike Jupiter or Saturn, Uranus is almost entirely featureless as seen from either the Earth or the Voyager spacecraft. Calculations indicate that the basic atmospheric structure of this planet should resemble that of Jupiter and Saturn, although on Uranus the upper condensation clouds are composed of methane rather than ammonia. However, Uranus (alone among the giant planets) does not have a substantial internal heat source. With little heat rising from the interior, atmospheric convection is suppressed to yield a stable atmosphere with little visible structure. In addition, our view is obscured by a deep, cold, hazy stratosphere.

In spite of the exaggerated seasons induced by the tilt of its axis, Uranus's basic winds blow parallel with its equator, just as on Jupiter and Saturn. The mass of the atmosphere and its capacity to store heat are so great that the alternating 40-year periods of sunlight and darkness have little effect; in fact, Voyager measurements show that the atmospheric temperatures are a few degrees higher on the dark winter side than on the hemisphere facing the Sun.

Deep Blue

Neptune, the final giant planet visited by Voyager, is so distant that it can be studied from Earth only with great difficulty, and the Voyager scientists really did not know what to expect. Disappointed by Uranus, they hoped for something more exciting at Neptune.

Whereas the discovery of Uranus was unexpected, Neptune was found as the result of mathematical prediction. By the first decade of the nineteenth century, astronomers were having difficulty accounting for the motion of Uranus. The discrepancies between its observed and predicted positions indicated that there must be an unknown planet pulling Uranus from its estimated path. Although most astronomers thought the data were insufficient to calculate the position and mass of the perturbing planet, the problem appealed to two relatively unknown young mathematicians, Urbain Jean Joseph Leverrier in France and John Couch Adams in England.

Both Adams and Leverrier obtained nearly identical results for the orbit and location of the hypothetical perturbing planet. Adams finished his calculations first and attempted unsuccessfully to deliver

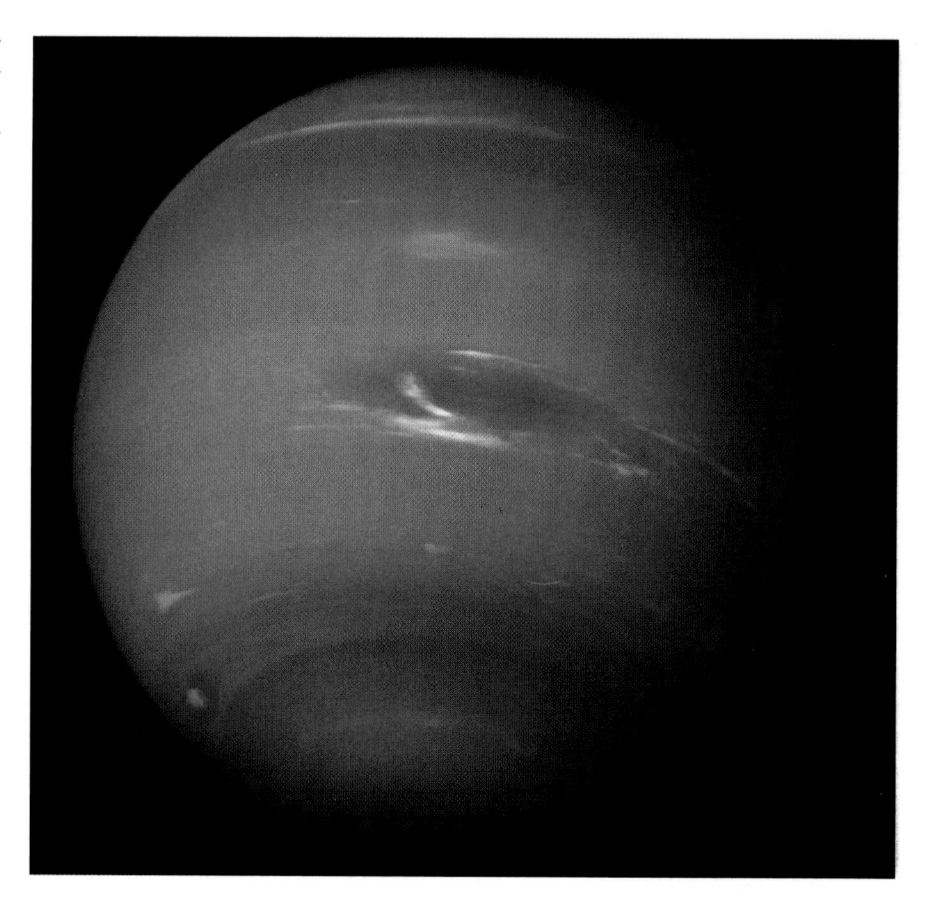

Neptune is more interesting than its sister planet Uranus, as shown in this Voyager image obtained in 1989. Atmospheric convection results in bright and dark clouds that can be seen even through the deep blue stratospheric haze of the planet.

them in person to the English Astronomer Royal in late September 1845. He tried again a month later but was rebuffed because the Astronomer Royal was at dinner, and his butler would not allow him to be disturbed. Meanwhile, Leverrier published the first part of his own calculations in France; during the spring of 1846, he completed his analysis. With the results published, the English finally initiated an effort to find the predicted planet, but they were too late. When Leverrier had had no success convincing the French astronomers that they should bother to look for the new planet, he wrote to a friend at the Berlin Observatory. The search began the very night Leverrier's letter was received, and it yielded instant success, with the new planet just where it had been predicted to be. Today the honor for the discovery of Neptune is justly shared by Adams and Leverrier (but not, note, by the astronomer who actually found it at the telescope of Berlin Observatory, one Johann Galle).

Neptune has a mass, diameter, rotation period, and composition very nearly the same as Uranus, but it has an internal energy source, like Jupiter and Saturn, as well as a more normal rotation axis. No one knows whether the strange axis and absence of internal energy on Uranus are related or just coincidental.

When the Voyager photos arrived at Earth, it was apparent that Neptune differed dramatically from Uranus in appearance. The upper clouds are composed of methane, but most of the atmosphere above this level is clear and transparent, with less haze than on Uranus. Scattering of sunlight lends Neptune a deep blue color similar to that of the Earth's atmosphere. Another cloud layer exists at a lower level, perhaps composed of hydrogen-sulfide ice particles. The primary difference between Uranus and Neptune, however, is the presence on

High clouds in the atmosphere of Neptune cast their shadows on the smooth cloud deck beneath. These clouds are composed of crystals of methane ice injected into the stratosphere by atmospheric convection.

Neptune of convection currents from the interior, powered by the planet's internal heat source. These currents carry warm gas above the regular methane cloud level to form additional high-altitude clouds, which produce bright white patterns against the blue planet beneath. These high clouds can even cast distinct shadows on the lower cloud tops, permitting their altitudes to be measured.

Neptune's weather is characterized by strong east-west winds generally similar to those observed on Jupiter and Saturn. The equatorial jet stream on Neptune actually approaches supersonic speeds. Neptune also has an atmospheric feature surprisingly similar to the jovian Great Red Spot. Called the Great Dark Spot, this storm is nearly 10,000 km long. Like Jupiter's Red Spot, it is found at latitude 20 degrees south, and its size and shape are similar relative to the size of the planet. Just as with the jovian Red Spot, we do not yet understand the origin of Neptune's Dark Spot.

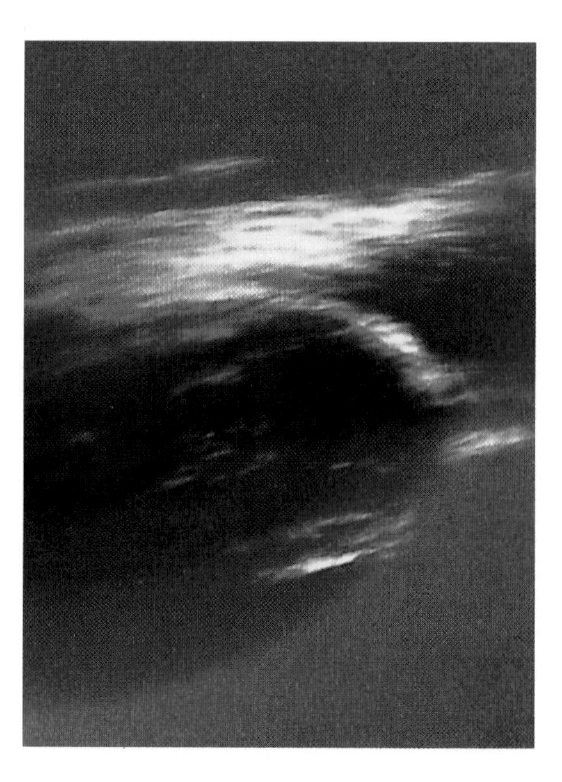

The Great Dark Spot of Neptune resembles Jupiter's Great Red Spot in its shape and location on the planet. Like the Red Spot, the Dark Spot is a high pressure region in the atmosphere. The spot is accompanied by streamers of bright methane cirrus clouds that form around it at a higher elevation.

Magnetic Fields and Magnetospheres

We end this chapter on the giant planets by briefly considering their magnetic fields and surrounding magnetospheres, the regions of space around planets where the planet's own magnetic field dominates and can trap electrons and ions and accelerate them to high energies. The Earth has a magnetosphere, which roughly coincides with the Van Allen radiation belts (see Chapter 3). Magnetospheres are the largest features of the giant planets.

To produce a magnetosphere, we must begin with a magnetic field. The magnetic fields of all four giant planets were probed by the Voyager spacecraft, and most of what we know about them results from analysis of that data. Planetary magnetic fields are produced by electrical currents in their interiors.

Jupiter's surface magnetic field is 25 times as strong as the Earth's. The jovian magnetic axis, like that of the Earth, is not aligned exactly with the axis of rotation of the planet, but tipped some 10 degrees. Nor does the magnetic axis pass exactly through the planet's center: it is offset by about 18,000 km. In addition, the jovian field has the opposite polarity of the Earth's current value;

Each of the giant planets has a substantial magnetic field, generated in its core. However, these magnetic fields are quite different when studied in detail; in particular, the fields of Uranus and Neptune are offset from the centers of the planets and cocked at large angles to the planets' axes of rotation.

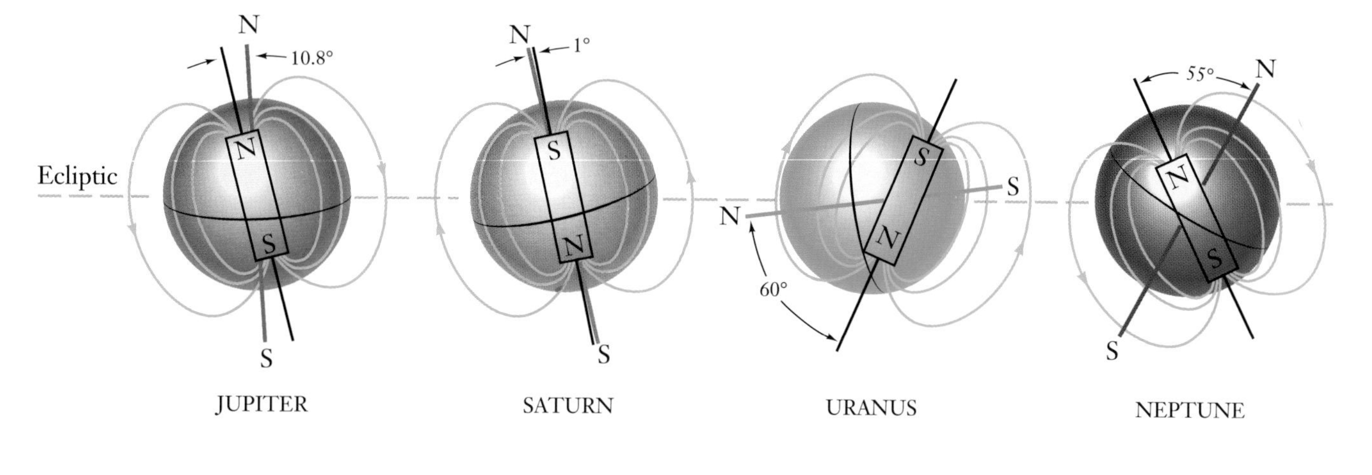

however, the Earth's field is known to reverse polarity from time to time, and the same may be true of Jupiter's.

Saturn has a smaller but substantial magnetic field, almost perfectly aligned with its rotation axis. The field of Uranus has a strength comparable to that of Saturn, but the orientation is very different. Like Jupiter's field, it is offset from the center of the planet, but to a greater degree (by about one-third of the planet's radius). In addition, the magnetic field of Uranus is tilted by 60 degrees with respect to the axis of rotation – the extreme opposite case from that of Saturn. Finally, Neptune's magnetic field is similar to that of Uranus, with its axis tilted by 55 degrees from the rotational axis and offset by nearly half the planet's radius.

Presumably the magnetic fields of the outer planets are generated in much the same way as the field of the Earth, by turbulence in an electrically conducting core. All of these planets spin rapidly, so there is a ready source of energy to power their internal dynamos. Jupiter and Saturn have large interior regions of metallic liquid hydrogen that act like the liquid iron core of the Earth. In the case of Uranus and Neptune, however, the metallic region may be in the hydrogen-water mantle, possibly accounting for the large offset of the field from the center of the planet.

The jovian magnetosphere is one of the largest features in the solar system. It is much larger than the Sun and completely envelops the innermost satellites of Jupiter. If the jovian magnetosphere were a visible entity, it would appear, "seen" from Earth, the size of our Moon. The total mass of the ions and electrons in this magnetosphere, however, is less than the mass of the Great Pyramid of Giza in Egypt. The primary sources of charged particles are ions chipped from the surfaces of the icy satellites of Jupiter or erupted from the volcanoes of its innermost large satellite, Io (see Chapter 5). The dominant ions are sulfur and oxygen, both products of Io's unique volcanic activity.

Ions and electrons within Jupiter's magnetosphere, accelerated by the spinning magnetic field of the planet, eventually reach extremely high energies. These energetic particles threaten spacecraft passing close to Jupiter, since they are capable of destroying sensitive electronic circuits; they would also endanger humans, if any should

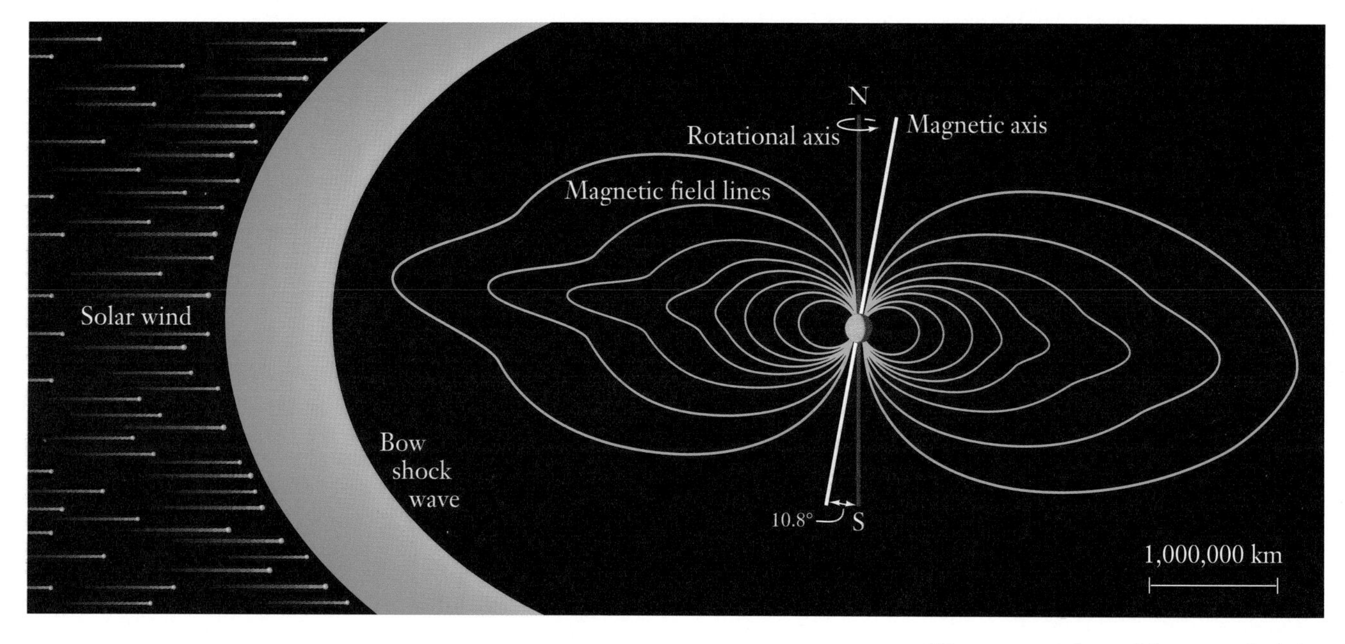

The magnetosphere of Jupiter, which has a diameter of several million kilometers, is the largest feature in the solar system. This huge cavity is filled with energetic electrons and ions, trapped by the magnetic field of the planet. Its boundary is defined by the balance between internal magnetic pressure and the pressure of the ions and electrons of the solar wind. The bow shock is the location where the solar wind is decelerated and begins to stream around the jovian magnetosphere.

ever venture close to the planet. On or near Io, an astronaut could survive for only a few minutes. Special shielding is required for spacecraft electronics, and no craft has been built that could last more than a few hours in this environment. Therefore, an Io lander or orbiter is far beyond our present capabilities, and direct human exploration of the inner jovian system may never be possible.

The magnetospheres of the other three jovian planets are generally similar to that of Jupiter. The physical dimensions of Saturn's magnetosphere are about one-third as great, and those of Uranus and Neptune still smaller, approximately in proportion to the sizes of the planets themselves. However, none of the other planets have volcanically active satellites to generate sulfur and oxygen ions, so their magnetospheres are less energetic and pose little danger to orbiting spacecraft. In this sense it was harder to build the Galileo Jupiter orbiter than the similar Cassini orbiter intended for Saturn.

We will return to the outer solar system later when we discuss the rings and satellites of the giant planets. Now, however, we turn our attention to the small, rocky worlds close to the Sun.

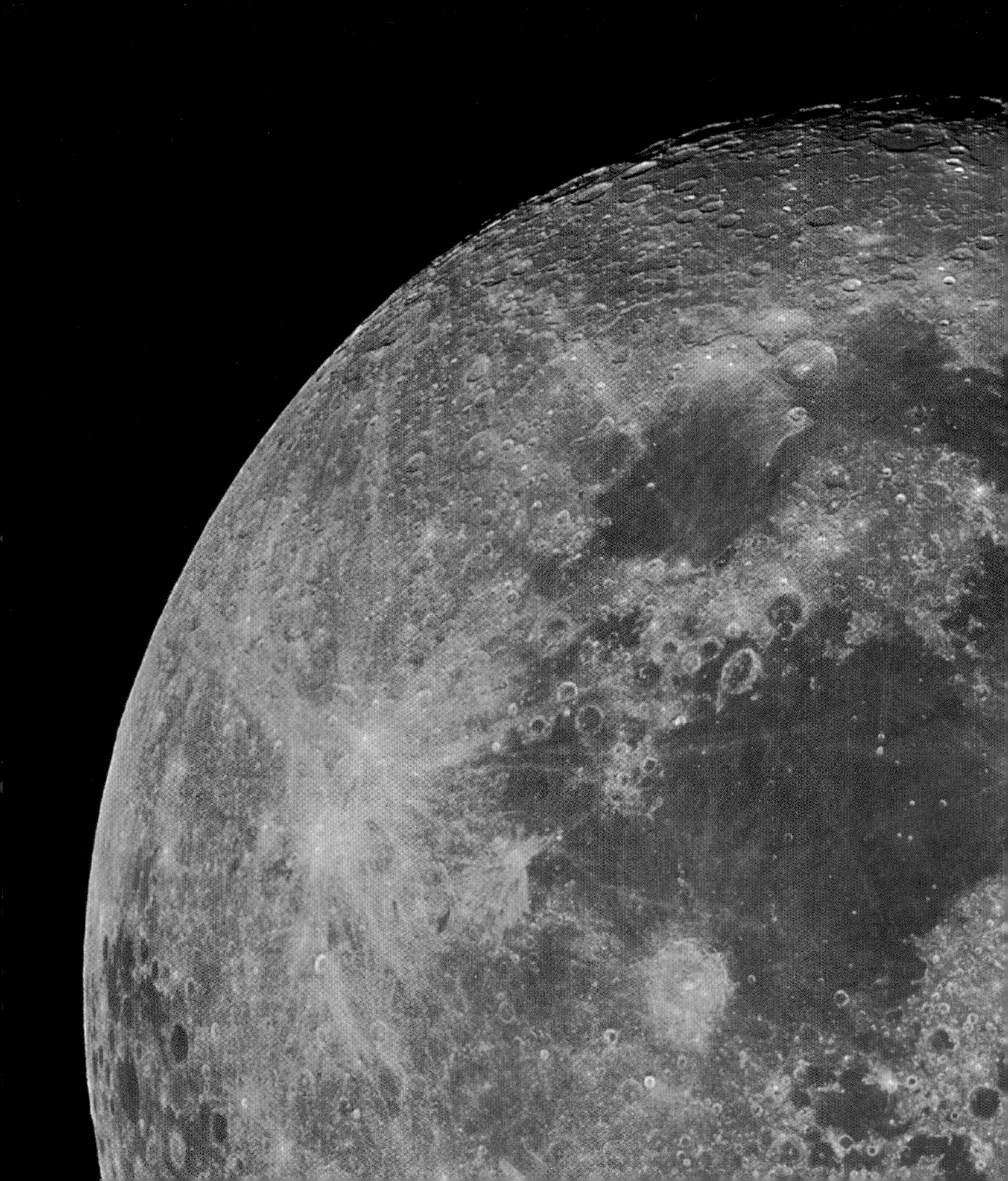

CRATERED WORLDS

The Moon and Mercury

The Moon is better known than any other planetary body beyond the Earth, thanks primarily to the Apollo exploration program of 1968–1973 and the lunar samples returned for laboratory analysis.

The Moon is an excellent place to begin our tour of the inner solar system. It is close to us and can be studied with even a small telescope. More important, we have surveyed it from orbit and sent robotic landers to its surface; it has even been visited by human explorers. We have hundreds of kilograms of samples of lunar material to analyze in our laboratories. With all of this information, we should be able to understand our own satellite better than any more distant world.

Another advantage of the Moon is its relative simplicity. As a general rule, larger planets are more complex, since internal heat can be retained by larger objects to power an active geology. Smaller bodies will be less geologically active (hence probably simpler to understand). With only a little more than 1 percent of the Earth's mass, the Moon should represent such a simple case.

Lack of a lunar atmosphere also considerably eases our task. There are no obscuring clouds or vapor, no wind or precipitation to erode the surface, and no oxygen or water to weather the rocks or alter their chemistry. Change on the Moon is a slow process; a million years from now, the footprints of Apollo astronauts will still be fresh. Unlike the Earth, where the surface is often hidden under vegetation and where rock and soil chemistry are altered by local conditions, on the Moon what you see is what you get.

The Moon also provides an appropriate introduction to the terrestrial planets because of the dominant influence of impacts in shaping its surface. All the terrestrial planets have been subject to a similar rain of debris, providing a common element in the histories of different worlds. By first examining impacts on objects like the Moon and Mercury, where such processes are dominant, we will gain the understanding needed to interpret more complex planets where both impacts and internally driven processes are important.

Another basic property of the Moon is its peculiar composition, relative to the Earth. From its density, we know that the bulk of the Moon is made of rock (silicates), with little or no metal core. Thus the Moon lacks the metallic component that we might have expected by analogy with the Earth. Analysis of lunar rocks yields another fundamental difference: the lunar material is extremely dry, lacking water and other volatiles even in chemical combination with the rocks. These distinctive aspects of the Moon's composition pose a challenge to any theory of the origin of our satellite.

Expeditions to the Moon

The Moon is the only planetary body that can be distinguished with the naked eye as a globe, and even without a telescope we can see that its surface is not uniform. Many cultures have associated names and myths with the markings familiarly known as the Man in the Moon, but it was not until Galileo's first telescopic studies that it became clear that the surface of our satellite was rugged and mountainous like that of the Earth.

Without a telescope, a sharp-eyed observer can see features on the Moon as small as one-fifteenth of its apparent diameter, or approximately 200 km across. But only a telescope, with its higher resolution, is able to reveal lunar surface topography and stimulate us to consider geologic evolution. For the Moon, the best telescopic resolution is about 1 km. Even after initial spacecraft exploration, most of the other bodies in the solar system have been imaged with resolutions of only 1 or 2 km, so our current photography of most other worlds is no better than that of the Moon before the space age began.

Galileo made the first telescopic observations of the Moon in 1610. In this drawing, the ragged sunrise line indicates the presence of topography and suggests that Galileo was able to identify a few of the largest craters on the Moon.

The Soviet Union led in the initial spacecraft exploration of the Moon, with the first flyby (1959), the first photography of the unexplored far side (1959), and the first successful transmission of images and other data from the lunar surface (1966). During the 1960s the United States pushed a parallel three-part effort at unmanned exploration in anticipation of the Apollo landings. The

Discovery of life on the Moon was announced to the American public by the New York Sun *in 1835. This fanciful illustration is supposed to have shown what astronomers were seeing with new and more powerful telescopes.*

Ranger spacecraft were crash-landers, designed to transmit a few very high-resolution images before impact. Five Lunar Orbiters mapped the surface and located potential landing sites. Finally, Surveyor landers tested the detailed physical and chemical properties of the lunar surface and certified the safety of the initial Apollo landing sites.

The 1960s were the years of the Great Space Race, with the United States and the U.S.S.R. struggling to be the first to land humans on the Moon. Rocket failures finally forced the Russians to abandon their effort, but Apollo astronaut Neil Armstrong succeeded with his historic first step on the Moon on July 20, 1969. Eleven others followed him in six successful lunar landings, each more ambitious and productive than its predecessors. By any standard, the Apollo Program was a supreme achievement. That we landed on the Moon at all was remarkable; that we did so in so short a time, and without losing a single astronaut in space, is little short of miraculous.

Apollo Missions to the Moon

Flight	Date	Landing Site	Accomplishments
8	Dec 1968	orbiter	first human circumlunar flight
10	May 1969	orbiter	first lunar orbit rendezvous
11	Jul 1969	Mare Tranquillitatis	first human landing
12	Nov 1969	Oceanus Procellarum	first ALSEP
13	Apr 1970	flyby	accident; landing aborted
14	Jan 1971	Mare Nubium	first mobile equipment cart
15	Jul 1971	Imbrium/Hadley	first rover
16	Apr 1972	Descartes	first highland landing
17	Dec 1972	Taurus Mountains	last flight; only geologist

For a few glorious years, it seemed that humanity had broken the bonds of Earth and truly begun a new space age. But at the peak of its success, Apollo lost its political appeal, and the program was abruptly terminated. The last human footprints were planted in the lunar soil on December 14, 1972, by Jack Schmitt, the only scientist-astronaut to reach the Moon. Today no nation has a capability for manned lunar exploration. The U.S.S.R. no longer exists, while in the United States, the giant Saturn rockets—destined for the Moon—rust on the grass at Cape Canaveral and Houston. Leftover Apollo spacecraft, built at costs of hundreds of millions of dollars, take the place of honor in museums instead of resting, as intended, on the lunar surface. No science-fiction writer had predicted that humans, having once attained the Moon, would so quickly abandon it.

Astronaut Buzz Aldrin stands on the lunar surface in this photograph taken by his Apollo-11 partner Neil Armstrong. The lunar surface consists of rocks and fine-grained dust produced from countless small impacts.

The scientific legacy of Apollo remains, however. Most important is the role of the lunar samples returned to the Earth. Sample collection was a primary scientific objective of every Apollo landing. The first thing each astronaut did upon alighting on the surface was to scoop up a sample of soil, assuring the return of some material even if the moonwalk had to be abandoned. In the later missions, rocks were carefully documented: measured and photographed in

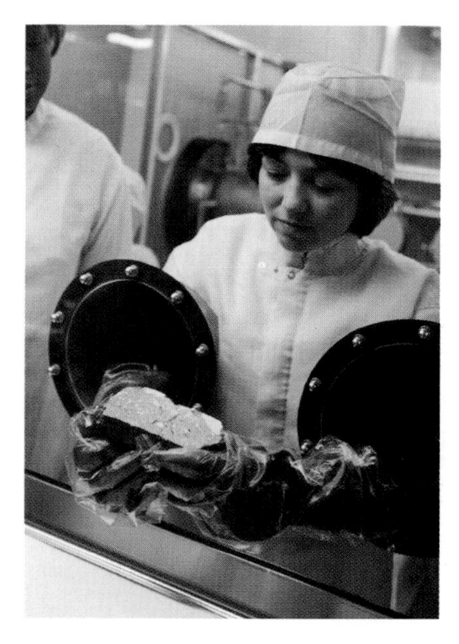

Many scientists believe that the most scientifically important legacy of the Apollo program consists of the returned samples, such as this large moonrock being examined at the NASA Johnson Space Center in Houston.

place before being picked up, then placed in individual, labeled bags for the trip to Earth. All samples were transported back to the Houston Lunar Receiving Laboratory in sealed containers, where they were inspected and cataloged under conditions that protected them from contamination or corrosion by oxygen or water vapor in the Earth's atmosphere. Most of the 382 kg of returned samples are still in Houston and are made available for analysis to scientists from all over the world.

Each Apollo landing after the first left behind an automated surface laboratory called the Apollo Lunar Surface Experiments Package (ALSEP), including a seismometer to measure moonquakes. Because the lunar environment is quieter than that of Earth – without winds, waves, truck traffic, and so on – these seismometers were much more sensitive than their terrestrial counterparts. In the later missions, the spent Saturn upper-stage rockets and the Lunar Landing Modules were deliberately crashed into the Moon to generate artificial moonquakes to be tracked by the ALSEP seismic instruments. Thanks to these experiments, we know more about the interior of the Moon than any other planetary object except the Earth itself.

A third important aspect of Apollo was scientific study of the Moon from the orbiting Command Modules. These spacecraft carried a number of instruments, including highly sophisticated mapping cameras on the last three missions. Other complex instrumentation made orbital maps of magnetic fields, chemical composition, and surface radioactivity as the Command Modules passed over the Moon.

Most of the information discussed in this chapter is derived from the Apollo flights or subsequent analysis of returned moonrocks. This wealth of data sets a standard against which to measure our progress in the exploration of other, more distant worlds.

The Face of the Moon

The Moon rotates on its axis once each orbit, so it keeps the same face turned toward the Earth; one hemisphere is always visible to us, while the other – the lunar far side – remained *terra incognita* until

Topographic detail is best seen when the Moon is near first or last quarter and sunlight illuminates the surface obliquely. Craters are especially prominent near the line separating the day and night hemispheres. In contrast, the Moon shows virtually no topographic detail at full phase, although intrinsic brightness differences, as between the mare and highlands, are easily seen. Note also the long bright rays emanating from young craters such as Tycho, near the bottom of the image.

revealed by spacecraft. In addition, the Moon experiences its familiar monthly cycle of phases as its angle of illumination by the Sun changes. If we watch the cycle of phases through a telescope, we see that lunar topography shows up best near first and last quarter, when sunlight strikes the surface obliquely and every hill or valley is sharply delineated.

What can we learn about the Moon from telescopic views with a typical surface resolution of 1 km? The most obvious conclusion is the presence of two distinct kinds of surface terrain. The predominant type is relatively light in color and extremly rugged, with craters of all sizes superimposed one upon the other. Since these

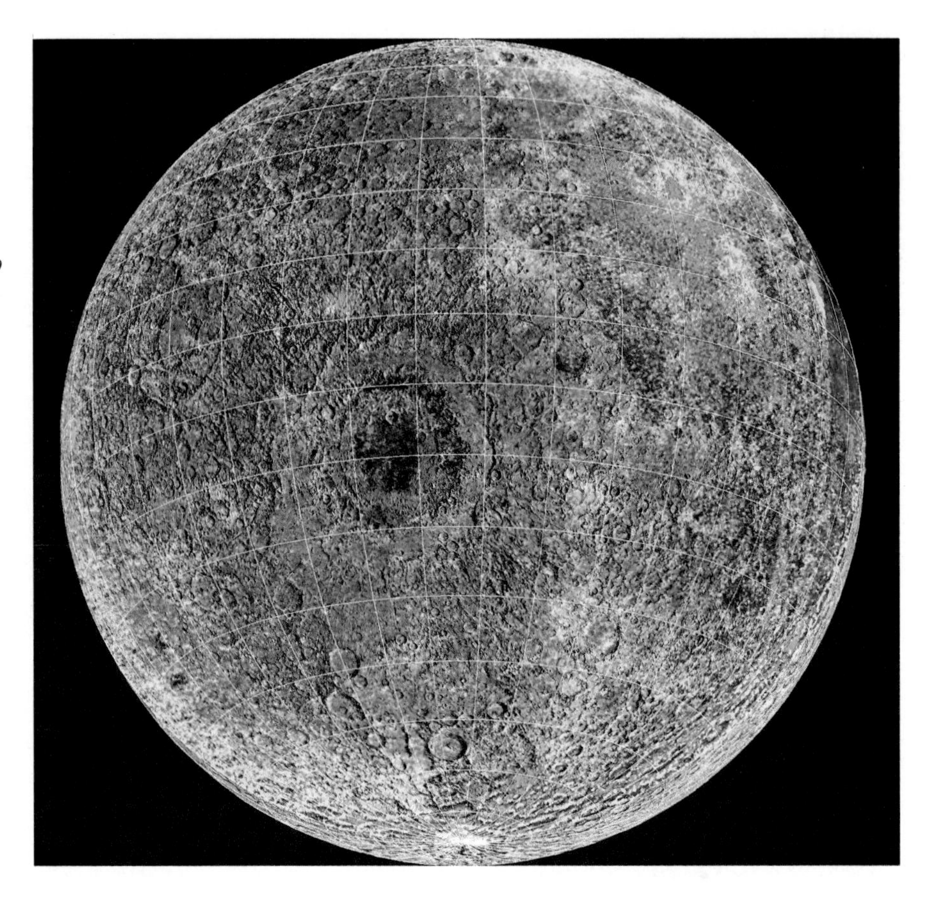

In 1991 the Jupiter-bound Galileo spacecraft flew past the Moon and obtained this view of the far side. With its advanced instrumentation, Galileo was able to measure subtle color differences that represent a variety of surface materials. These color differences are greatly exaggerated in this image; to the human eye the Moon appears a uniform grey.

heavily cratered regions also generally lie at higher elevations, they are called the lunar highlands. The second type is darker and smoother, with few large craters. These regions, which create the features of the Man in the Moon, are called maria. Mare (plural, maria) is the Latin word for sea, and when the term was first applied to the Moon these darker regions were thought to be water oceans. Only 17 percent of the total lunar surface consists of mare material.

Maria and highlands have different color and reflectivity, implying that their chemical compositions differ. The disparity in cratering also suggests distinctions in age and geologic history. Finally, the range of elevation is, as we will see, related to the largest-scale forces that have molded the lunar surface. The two types of

This view of Mare Imbrium includes numerous secondary craters and other ejecta from the large crater Copernicus, on the upper horizon. Copernicus is almost 100 km in diameter. Streams of ejected material produce the bright rays visible at full phase through even a small telescope.

terrain provide windows into two different periods of solar system history.

The prevalence of circular craters is one of the most striking features of the lunar surface. The word crater (derived from the Greek for cup or bowl) refers to the shape of the feature. Note that a crater is a depression; a volcanic mountain, such as Vesuvius in Italy or Haleakala in Hawaii, frequently has a crater at its summit, but the mountain itself is not a crater. Approximately 30,000 craters greater than 1 km in diameter are found on the Moon.

Lunar craters range from sharp, new-looking depressions of nearly perfect bowl shape to old, battered craters that can barely be distinguished, so obscured are they by later impacts. In the highlands,

King Crater on the far side of the Moon is a relatively fresh impact structure 75 km in diameter. It shows most of the features associated with large lunar impact craters, including a flat floor, central peaks, terraced walls, and a hilly apron of ejecta surrounding the crater itself.

the craters are packed shoulder to shoulder, so that new additions must occur at the expense of those craters already present. In some highland areas, individual craters almost become lost in the accumulation of jumbled, moutainous debris. In contrast, the craters on the maria are rather widely spread, suggesting that there has been less time for them to accumulate.

Viewed on the largest scale, the face of the Moon is dominated by a few ringed impact basins, essentially very large craters. On the side of the Moon facing the Earth, these basins tend to be the sites of the darker maria. The major lunar mountain ranges define the margins of these basins. Unlike the mountains of Earth, these mountain arcs were thrown up as part of the basin-forming process, which we will now examine.

Engines of Geologic Change

The lunar surface represents a balance between two different geologic processes. The primary external influence is the formation of craters by the impact of asteroids and comets. The most important internal activity, the eruption of lava onto the surface, destroys preexisting craters and produces broad volcanic plains.

The lunar highlands are the oldest surviving part of the Moon's crust. From their heavily cratered surfaces we conclude that they were formed in another era, at a time when there was much more debris striking the surfaces of the inner planets. An indication of the magnitude of highland cratering is provided by the behavior of moonquakes studied by the ALSEP instruments, which showed that the crust has been shattered to a depth of 25 km. The highland rocks themselves are made up of fragments broken and scattered by previous impacts.

Thanks to the samples returned from the Moon, we can date many events of lunar history with precision. Like terrestrial geologists, lunar scientists can use the residual radioactivity of lunar rocks to determine the amount of time elapsed since these rocks congealed from a molten state. Radioactive age determinations for highland rocks suggest that most of them formed about 4 billion years ago; the oldest highland fragments returned by the Apollo astronauts have ages of 4.4 billion years. In contrast, we will see that

The old, heavily cratered highlands make up 83 percent of the Moon's surface. The craters, which are packed shoulder to shoulder, date back to the period of heavy bombardment that ended about 4 billion years ago. The largest crater in this Apollo 11 photograph is 50 km in diameter.

the lunar maria are younger, although still ancient by terrestrial standards.

The large lunar impact basins were formed during the final stages of this heavy bombardment. Today lunar scientists have identified about 30 ancient basins, including many on the lunar far side. The youngest of the great basins are Imbrium, on the side of the Moon that faces the Earth, and Orientale, which cannot be seen completely except from a spacecraft. Both Imbrium and Orientale are mountain-ringed, circular features about the size of Texas.

Even before the end of the period of heavy bombardment, volcanic vents were erupting on the lunar surface. However, the major period of lunar volcanism apparently did not begin until shortly after the formation of the Imbrium and Orientale basins. Over about a half-billion years, repeated outpourings gradually filled the near-side basins to create the familiar pattern of lunar maria. The mare lavas are composed of basalt, like similar eruptions on Earth. These basalts range in age from 3.9 to 3.2 billion years, when large-scale volcanism ceased.

There are no large volcanic mountains on the Moon, but eruptions there produced other interesting geological features like lava valleys. Before Apollo, many scientists thought these long, curving channels, which resemble terrestrial rivers, had been carved by running water. We now know that the Moon is, and always has been, dry. Since these valleys are found in the maria, it seems likely that they, too, represent some kind of volcanic phenomenon. (In the next chapter we will discuss even longer lava channels recently discovered on Venus.)

Substantial information on the source of the lunar eruptions has been derived from a detailed study of the composition of the lunar basalts, and the results provide insight into the history of the Moon. Basically, the conclusion from this work is that the mare samples represent material that has been three times chemically separated (or fractionated) from the original material of the solar nebula. There was initially loss of water and other volatiles from the entire Moon. Second came differentiation, the process in which the Moon melted and its interior separated into layers of different composition and density. Finally, in a process called partial melting, the more easily melted minerals of the mantle were extruded as lava. This partial

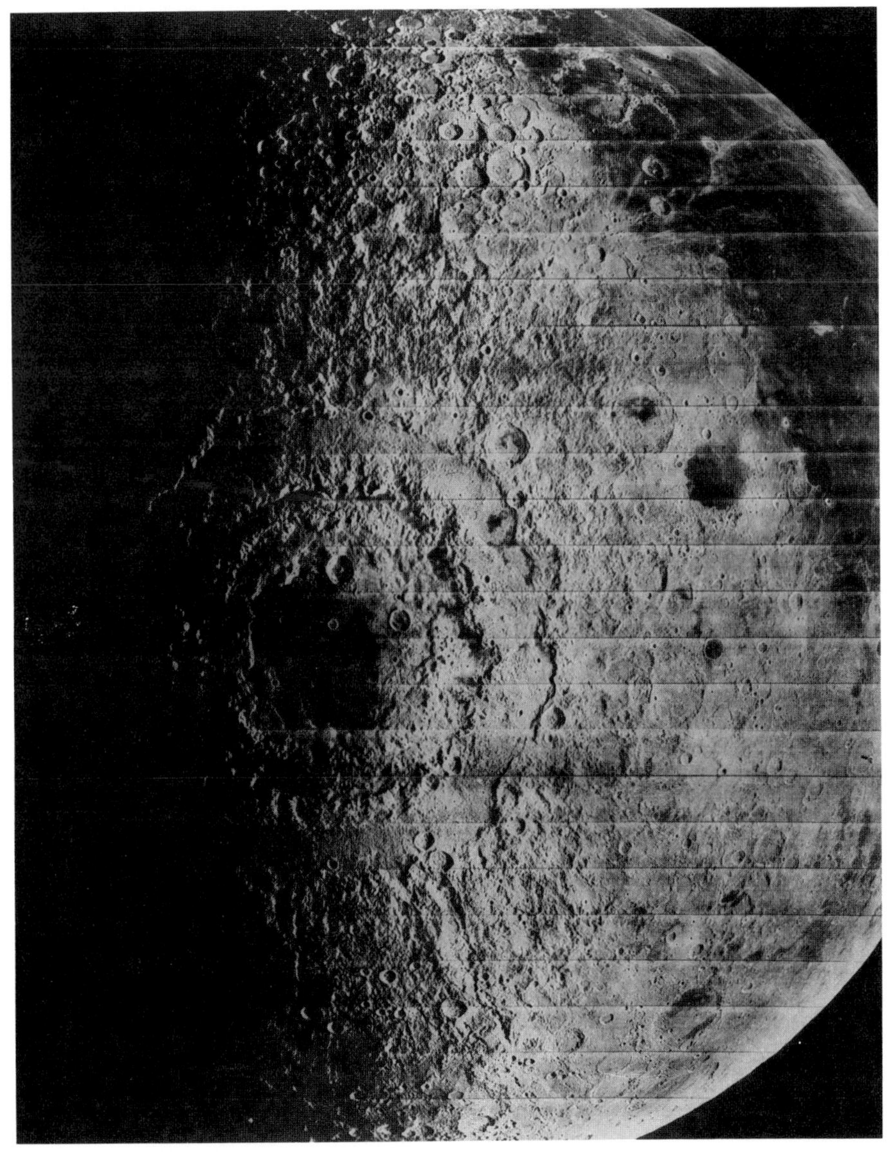

The youngest of the large impact basins is Orientale, which formed 3.8 billion years ago. Unlike most of the other basins, this 1000-km-diameter scar has not been filled in with lava, so it retains its striking "bull's eye" appearance.

melting took place at a depth of several hundred kilometers.

To summarize, we see four major periods of lunar history. First the Moon differentiated into a layered structure, with the lightest minerals floating to the top to form the crust. During the half-billion-year period that followed, impacts from external sources produced the heavily cratered highlands. Shortly after the end of this heavy bombardment, partial melting in the lunar mantle gave rise to a long period of lunar volcanism, forming the dark maria. Since the cessation of both heavy bombardment and lunar volcanism, the surface has continued to absorb a slow rain of impacts from space.

Impact Craters

Craters are the dominant geologic features on the Moon, readily visible to generations of telescopic observers. Yet their origin in the impacts of cosmic debris was not widely recognized until about 50 years ago. It is interesting to examine what finally provided convincing evidence of an impact origin for the lunar craters.

From the invention of the telescope until well into the twentieth century, most astronomers and geologists thought that the lunar craters were volcanic. The argument was simple. No impact craters had been recognized on Earth, while volcanoes were well known to terrestrial geologists – and volcanoes do have craters. Therefore the craters of the Moon, by analogy with the Earth, must be volcanic.

The first detailed arguments against the volcanic crater hypothesis were presented in the 1890s by Grove K. Gilbert, then director of the U.S. Geological Survey. Assembling data on the sizes and shapes of lunar craters and of terrestrial volcanoes, Gilbert pointed out the differences between the two. Most significant was the fact that lunar craters do not appear at mountain summits; indeed, the floors of most lunar craters actually lie well below the level of the surrounding plains. Gilbert developed a strong case for the dissimilarity between lunar and terrestrial craters. The argument by analogy crumbled, and Gilbert postulated that the craters were instead the result of impacts.

General acceptance of this hypothesis did not come until scientists gained a better understanding of the impact process itself,

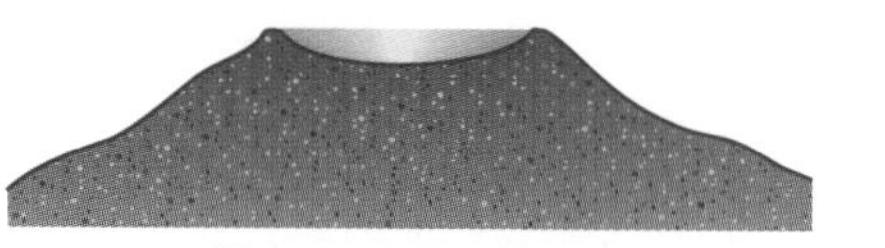

Terrestrial volcano

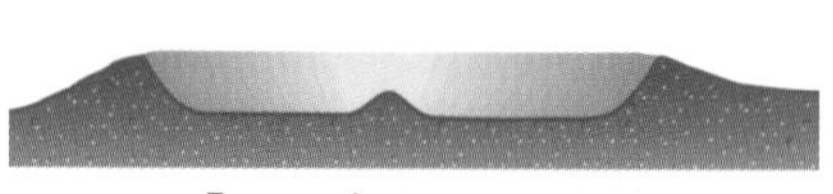

Lunar impact crater

Terrestrial volcanic craters and lunar impact craters have distinctly different shapes, as pointed out a century ago by G.K. Gilbert. Volcanic craters are depressions at the summits of volcanic mountains. In contrast, the large lunar craters are flat-bottomed ringed plains, with floors generally lower than the adjacent surface.

which is far different from that by which a tossed stone makes a depression in sand. Because they strike at great speed, projectiles from space disintegrate upon impact to produce violent shock waves and a fireball of superheated vapor, almost like the explosion of a large bomb. This explosion, not just the mechanical force of the impact, digs the crater. Just as bomb or shell craters are circular, even if the incoming projectile strikes at an angle, impact craters are nearly circular, independent of the direction from which the projectile struck.

Imagine an asteroid or comet crashing into the Moon at a speed of many kilometers per second. Its energy is so great that it penetrates two or three times its own diameter below the surface before it stops. The force of the blow shatters the surface and generates seismic waves — moonquakes — that rapidly spread throughout the Moon; these shock waves pulverize the surface rock and begin to scoop out the crater. Meanwhile, much of the impact energy goes into heating the projectile and its immediate surroundings. The material forms a pocket of superheated gas, and the expansion of this hot, high-pressure gas contributes to the formation of the crater. The vaporized remnant of the original projectile, plus several hundred times its mass in excavated rock, is thrown upward and outward. Some falls back into the crater, partially filling it; the rest is ejected, including large fragments that rain down on the surrounding lunar surface at distances of up to several thousand kilometers from the point of impact. The larger fragments form additional pits called secondary craters.

Large craters tend to have flat floors, often later flooded by lava. Imagine yourself landing in the interior of a lunar crater like Tycho or Copernicus, each of which is nearly 100 km in diameter. You would not have the sensation of being in a bowl-shaped depression; rather, you would find yourself on a level, rocky plain, with the distant rim visible as a low line of mountains along the horizon. In contrast, small, young craters really do look bowl-shaped, as we know from the experiences of the Apollo astronauts.

Only a handful of impact craters on Earth are recognizable from ground level. The best known of these is Meteor Crater in northern Arizona, formed 50,000 years ago. Meteor Crater is only about 1 km across, smaller than the smallest lunar crater visible through a

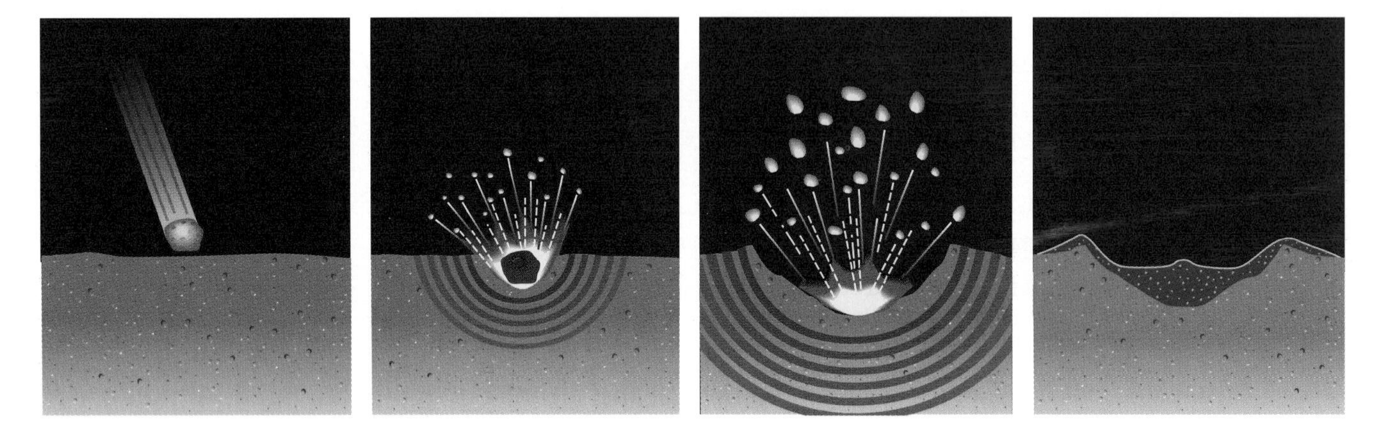

Most craters on the surfaces of the Moon and planets are the product of hyper-velocity impacts. Striking at a speed of many kilometers per second, the projectile is vaporized by the impact. Shock waves pulverize the surface rocks, while the central explosion digs a deep hollow and ejects huge quantities of debris—typically about 100 times the mass of the projectile itself. Part of this debris falls back into the crater, but substantial quantities are spread over the surrounding surface. Some ejecta escapes entirely, especially if the surface gravity is low.

telescope, but it gives us a good idea of what a fresh lunar crater would look like. Visit it if you can; it is the best example on our Earth of the most characteristic class of features on the terrestrial planets!

Lunar Chronology

The study of craters plays an important part in planetary science. The number of impact craters on a planetary surface is a measure of the age of that surface, where we define the age as the time over which the surface has been sufficiently stable to preserve craters once they are formed. On an active planet like the Earth, many erosional and geologic processes degrade and destroy craters. But on a small, airless world like the Moon, practically the only events that can destroy craters are later impacts or lava flooding during periods of large-scale volcanism.

When we see differently cratered regions on the Moon, we interpret the disparity as the result of surface age. The situation resembles that of a city street in the midst of a long, steady snowstorm. As you walk along, you will find the sidewalk in front of some houses deeply covered with snow, while in other places the depth is less, with a few areas of sidewalk nearly clear. Do you conclude that different amounts of snow have fallen in front of the Jones' house and at the Smiths' next door? No; you attribute the variable depth to the time passed since each section of walk was shoveled. The less snow, the shorter the time that has passed. It is the same with craters.

We can agree that the more craters there are on a surface, the older that surface is. But how much older? Only for the Moon, where scientists have measured both the crater numbers and the associated radioactive ages from returned samples, can this question be answered with any precision. On Earth there are too few craters, while on other planets there is no absolute age scale defined by returned samples.

Look at any picture of the lunar maria and you gain an impression of the number of craters that accumulates in a little more than 3 billion years, the age of most mare surfaces – that is, the time since they were last flooded by fresh lava. As far as we can tell, the rate of impacts has been fairly constant over the past 3 billion years. The mare cratering thus provides a benchmark against which to compare other planets. For example, if we see a plain on Venus with

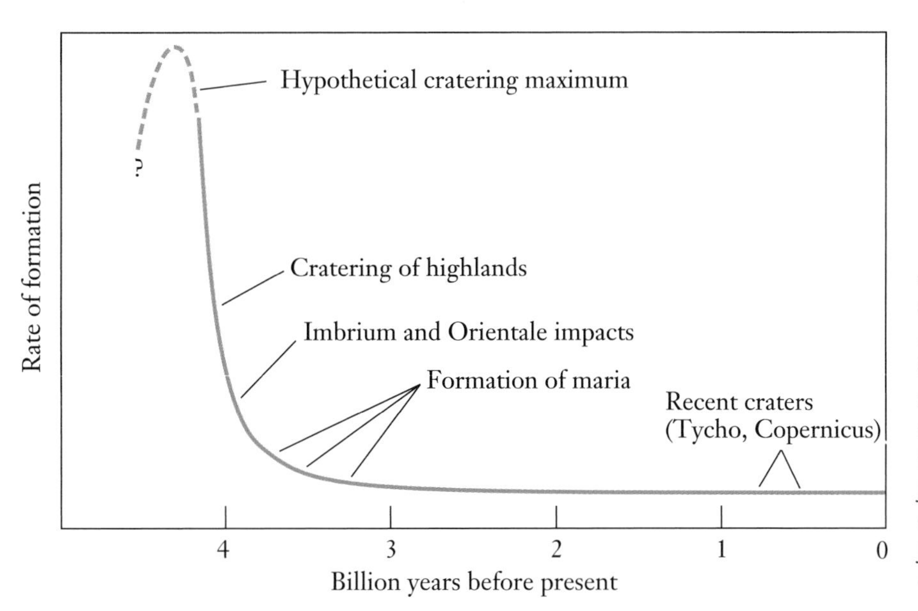

While the rate of bombardment of the Moon by asteroids and comets seems to have been relatively constant for the past 3 billion years, we know that the impact rate was previously much higher. This schematic cratering history was obtained by comparing observed crater densities on the lunar surface with dates derived from returned moonrocks. With such a curve, we can obtain approximate surface ages just by counting the number of craters in different parts of the Moon.

only one-fifth as many craters as on the lunar maria, we conclude that it is only one-fifth as old, or about 700 million years.

The lunar highlands have 10 times more craters than the maria. If the impact rate has remained the same, the highlands must be 10 times older than the maria, or about 35 billion years. This is impossible, however, since the Moon itself is only 4.5 billion years old. Our assumption of a constant cratering rate must be wrong. We can understand the heavy cratering of the highlands only if the impact rate was much higher prior to the formation of the maria. This is the reasoning behind the concept of a heavy bombardment early in lunar history.

With the aid of returned highland samples, we can trace at least the final stages of this heavy bombardment. It appears that the impact rate was a thousand times greater at 4.0 billion years ago than at 3.8 billion years. Earlier it may have been higher still, but the evidence concerning early lunar history is limited. This period around 4 billion years ago is variously called the late heavy bombardment or the terminal bombardment of the Moon. Because the impact rates were so high, most scientists suspect they were the result of a unique event, such as the disruption of a large asteroid that scattered fragments throughout the inner solar system. Alternatively, the terminal bombardment may represent the final stages of planetary accretion. Either way, this period of heavy cratering has largely obliterated direct evidence of the first half-billion years of lunar history.

The Lunar Surface

When the first Apollo astronauts stepped onto the lunar surface, they found themselves in a stark but beautiful world. The airless sky was deep black, the surrounding plains a dark brownish gray. The force of gravity was just one-sixth of that at the surface of the Earth, making their bulky spacesuits seem relatively light. With no haze to obscure the view, distant details stood out as sharply as those in the foreground. The mare plain itself was flat and covered with scattered rocks of all sizes and shapes. Some distance away the low profiles of small craters could just be identified, but at first sight there was little to establish that the observers stood on a heavily cratered planet.

Once they began to move about, the astronauts quickly became aware that the surface of the Moon was layered with a fine, dark dust that soon coated their spacesuits and eventually found its way into every item of their equipment. In spite of this omnipresent dry dust, they found the lunar soil to be firm underfoot. The bootprints they made were crisp and sharp-edged, as if made in damp dirt or crunchy snow.

Although later Apollo flights landed in rougher and more scenic locales than that first mare site, immediate surroundings were about the same everywhere. The entire Moon is covered with fragmented rock and dust, ejecta from impact craters near and far. This fine-grained soil accumulates at an average rate of about two millimeters per million years, or two meters per billion years. In addition to this

Apollo 17 astronaut Jack Schmitt inspects a large boulder in the Litrow Valley at the edge of the lunar highlands. The mountains of the Moon have smooth, rounded contours, primarily because there has been no water or ice erosion to sculpt them into steep peaks and valleys.

buildup, the upper part of the soil is frequently disturbed by very small impacts, which stir the material and help to maintain the loose, fine dust.

When examined under a microscope, lunar soils are found to contain large quantities of glass spherules, each the size of a grain of sand. Glass is silicate material that has been melted and then cooled very rapidly — in this case, lunar rock melted by meteorite impact. Most of the glass-forming impacts are very small. On Earth, the atmosphere protects us from these micrometeorites, so our soil does not contain similar glass spherules except at a very few sites associated with terrestrial impact craters. Most of the lunar glass is dark, and the presence of these tiny spherules contributes to the dark color of the Moon's surface.

A number of terrestrial volcanic landscapes are casually (and erroneously) referred to as "lunar." For example, the Craters of the Moon National Monument in Idaho is an area of fresh lava flows and cinder cones without a single impact crater. We know of no such landscapes on the Moon. Instead of young volcanic areas, we might better apply the metaphor "lunar" to dusty desert plains or to Meteor Crater in Arizona.

In the absence of moderating air or oceans, the surface of the Moon experiences much greater temperature extremes than the Earth. The daily temperature contrasts are further increased by the fact that day and night are each two weeks long. At the near-equatorial Apollo sites, the maximum surface temperature is about 110 C, higher than the boiling point of water; during the long lunar night, it drops to −170 C, only about 100 degrees above absolute zero. It is a tribute to the design of the Surveyor spacecraft and the ALSEP instruments that they were able to operate over this huge temperature range. Astronauts were never subjected to such extremes, since they always landed in the lunar morning when the temperatures were similar to those on the Earth.

The last three Apollo flights visited mountainous areas: the Apennines, the Descartes highlands, and the Taurus Mountains. Notable in each of these landscapes were the gentle, rounded contours of the lunar mountains and hills. Science fiction and space artists had always depicted the mountains of the Moon as steep and spiky, but the reality was otherwise. A partial explanation for the soft

Before we visited the lunar surface, the landscape was often depicted as consisting of very steep, spiky mountains—just the opposite of the real situation. This illustration is from the 1950 film Desination Moon, *which accurately represented mid-century ideas about the Moon and space travel. Compare this image with the photograph on page 69.*

contours can be found in the ejecta blanketing all lunar features. But the primary reason for the absence of sharp peaks or steep cliffs on the Moon is that there is no water or ice erosion, as on Earth, to cut deep valleys and shape mountain crags. In the absence of such forces, the mountains on any planet will be as gentle as those of the Moon.

Mercury

The one planet in our solar system that closely resembles the Moon is Mercury, which is also small, heavily cratered, and airless. Let us now examine Mercury and compare its history with that of the Moon.

Mercury is a disappointment to any telescopic viewer. Given its size and the difficulties of observing it near the Sun, only a persistent (and lucky) astronomer succeeds in seeing markings on its surface. Using radar observations, scientists in the 1960s finally measured the rotation period of the planet, which is 59 days, exactly two-thirds its 88-day orbital period.

Mariner 10 obtained this view as it approached Mercury in 1974. The surface is heavily cratered and looks remarkably similar to our Moon.

One thing that astronomers have known for a long time is that Mercury is extremely dense. By measuring its diameter and mass, we calculated a density of 5.4 g/cm^3, much greater than that of ordinary rock. The inescapable conclusion is that Mercury contains a large core of dense metal, probably mostly iron. In this respect at least, Mercury is different from the Moon, which is depleted of metal.

Although the interior compositions of the Moon and Mercury are very different, their surface material may be similar. While we have no samples from Mercury to study in the laboratory, the general colors and reflectivity of this planet match those of the Moon, indicating the presence of silicate rock. Whether there are basaltic lavas such as those of the lunar maria remains unknown, however. Similarly, there is no way to know for certain if the crust of Mercury is dry like that of the Moon, if it shares any of the other chemical peculiarities of our satellite, or even the age of the surface. In the absence of returned samples, interpreting the composition and history of another world is very difficult.

We have observed that Mercury has a fine-grained soil, like that of the Moon. Surface temperatures have also been measured, ranging from a high of about 400 C to a low of about −170 C over most of the night-side hemisphere. The lowest temperatures on Mercury are just about the same as those on the Moon, since the higher daytime temperatures are compensated by a longer cooling period during the 88-day mercurian night.

Almost everything scientists know about the geology of Mercury was learned from a single spacecraft, Mariner 10, which made three flybys of the planet in 1973 and 1974. The Mariner 10 cameras revealed a planet that looked remarkably Moon-like. Most geologists had anticipated greater evidence of internal geologic activity, but the Mariner photos, revealing only one lava-flooded impact basin, generally did not support the idea of widespread mare-like volcanism on Mercury.

The ubiquitous impact craters on Mercury bear many resemblances to their lunar counterparts, varying in numbers from values near that of the lunar maria up to several times as many. The lower values suggest that Mercury, like the Moon, has experienced some destruction of craters since the period of heavy bombardment. But lacking evidence that this planet has also undergone a period of

The Mariner 10 spacecraft followed a complicated orbital trajectory to its destination. Because there was insufficient rocket power to propel the spacecraft directly to Mercury, it first flew to Venus, where the gravity of that planet helped accelerate it inward to Mercury. Mariner 10 achieved an orbit with exactly twice the orbital period of Mercury itself, allowing the spacecraft to make three flybys separated by six-month intervals.

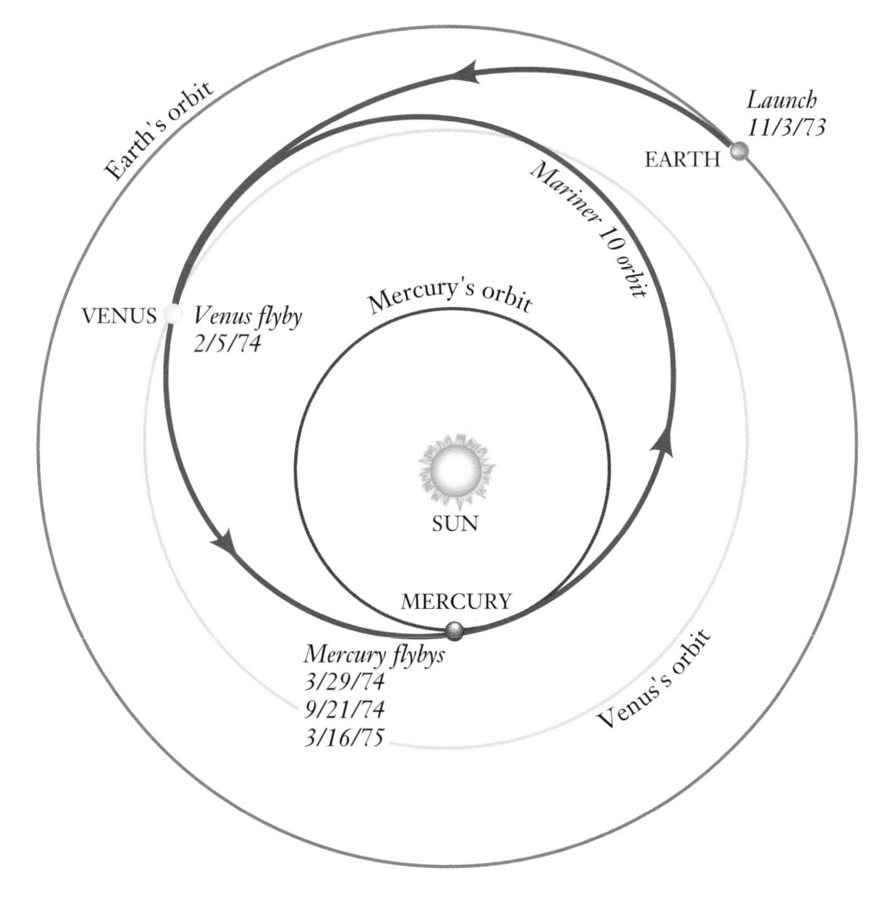

This photograph shows part of the Caloris Basin, which is 1300 km in diameter. This ancient impact basin is the largest structural feature on Mercury seen by Mariner 10. Compare this photograph with that of the Orientale Basin on the Moon (page 63) to see their great similarity.

volcanism that produced a counterpart to the lunar maria, we are unable to develop a history for Mercury that could draw on our more detailed knowledge of the Moon.

The less-cratered regions of Mercury simply do not look like lunar maria. Instead of dark basalts, these plains have the same brightness or color as their heavily cratered surroundings. Nor are they as flat as maria; rather, they are gently rolling plains. Thus geologists are left with a fundamental question: What destroyed some older craters to produce the plains? Was it volcanism that took place during the later stages of heavy meteorite bombardment but did not leave the characteristic marks of the lunar maria? Or was some other process responsible, such as blanketing of the surface by impact-produced ejecta? No one knows, although a vote among geologists would favor volcanism of some kind.

Geologically, the most remarkable features on Mercury are compressional cliffs or faults, just the sort of wrinkles that might form in the crust if the interior of the planet shrank slightly. Since they formed after most of the craters, they must represent an internal event on Mercury that took place many hundreds of millions of years after the solidification of the crust. No similar wrinkles or other evidence of shrinking exist on the Moon. Probably it was solidification of Mercury's metallic core that caused this global shrinkage.

In the two decades since the Mariner 10 flybys, we have learned more about Mercury from Earth-based observations. Among the interesting recent discoveries are the presence of an exceedingly tenuous atmosphere of the metal sodium and evidence, from radar studies, of ice near Mercury's poles, perhaps in permanently shadowed craters.

Planetary interiors are mysterious places, forever inaccessible to direct measurement. What little is known about them must be indirectly deduced and is inevitably subject to misinterpretation. Yet the effort must be made if we are to understand planetary evolution. Calculations show that Mercury contains roughly 60 percent metal by mass. If 60 percent of Mercury consists of an iron-nickel core similar to that of the Earth, this core must have a diameter of 3500 km and extend to within about 700 km of the surface. The planet is thus either enriched in metal or depleted of rock. Distinguishing between these alternatives requires us to understand the origin of Mercury. But first let us consider the origin of the Moon.

Mercury has several features produced by internal shrinkage of the planet. Discovery Scarp, shown in this Mariner 10 view, is nearly 1 km high and more than 100 km long. It cuts across several craters, proving that the compression of the crust took place after these craters were formed.

Origin of the Moon

In some ways the Moon is very much like the Earth, in other ways very different. Reconciling all the facts in a self-consistent theory of the origin of the Moon has proved difficult. Before lunar exploration began, "What is the origin of the Moon?" was a major question for planetary scientists. The Apollo program was supposed to answer this question, but it has required nearly 25 years since Apollo for a plausible explanation to emerge.

Study of lunar samples reveals a fundamental similarity in the detailed composition of rocks from the Earth and the Moon. Not only are many of the lunar minerals like those found on our own planet, but, in addition, the relative proportions of different isotopes of oxygen and other elements are the same. This similarity strongly suggests that the two bodies were formed together, perhaps out of the same mix of materials condensing from the solar nebula.

Unfortunately for this hypothesis, the Moon differs from the Earth in its bulk composition, which resembles that of the terrestrial mantle rather than that of the Earth as a whole. In addition, water and other volatiles are severely depleted on the Moon relative to their expected cosmic abundances. If the Moon and Earth formed together, it is difficult to explain the former's lack of metal and of volatiles.

Three traditional theories address the origin of the Moon, sometimes called the "daughter theory," the "sister theory," and the "capture theory." Let us confront each with the evidence just cited.

The daughter, or fission, theory supposes that the Moon formed from the Earth. In 1880 astronomer George Darwin, whose father was the biologist Charles Darwin, calculated that a rapidly spinning Earth could have split to form a sort of double planet. If this split took place after the differentiation of the Earth, the Moon might be formed from mantle material only, explaining why it is depleted of metals. Flaws in the daughter theory are associated with the detailed differences in composition between the Earth's mantle and the Moon. In addition, modern calculations do not support the idea that the Earth could spontaneously divide into two pieces.

The sister theory suggests that the Earth and Moon formed close together from a spinning cloud of dust, as many scientists believe the large satellite systems of Jupiter, Saturn, and Uranus originated. The problem here is simple: the sister theory provides no explanation for the large compositional differences between Earth and Moon. The daughter theory at least suggested how the Moon might have formed without much metal, but sisterhood does not seem to be compatible with the chemical evidence.

The third possibility is capture. The idea here is that the Moon formed elsewhere in the solar system and was subsequently captured into orbit around the Earth. Compositional differences are thus understood as representing different condensation conditions in separate locales in the solar nebula, but again there are problems. First, the isotopic data seem to be telling us that the Earth and Moon formed from the same pool of material. Second, we know of no way for the Earth to have captured the Moon into a stable orbit. Today, the capture theory has few defenders.

If the Moon is neither daughter, sister, nor interloper, what is it? Many scientists feel that the basic theories described above are too simple and that we must seek a more complex scenario. Perhaps elements of more than one of these theories were involved. We require a mechanism that permits the Moon (or its precursor materials) to form initially in the same part of the solar nebula as the Earth and then to undergo some process or processes that removed most of its metals and volatile elements before solidification. A potential solution to this problem is provided if the Moon formed as the product of a planet-shattering impact on the early Earth.

Worlds in Collision

Recently planetary and Earth scientists have become more aware of the role of impact catastrophes in solar system history. Even today asteroids collide and break apart, and occasional random impacts are capable of redirecting the course of life on our own planet – a subject we return to in the Epilogue. However, the primary focus of this new catastrophism is on the formative stages of the solar system. Calculations suggest that there was a time when there were not just the four inner planets we see today, but several dozen protoplanets of lunar mass or larger. Collisions among these protoplanets played an important role in early planetary evolution.

Imagine the consequences if the Earth were to have been struck by a protoplanet shortly after its formation. Suppose further that the Earth had already differentiated, with most of its metal sinking to the center to form a core. If the colliding object were the size of the Moon, it would produce an immense crater of nearly planetary dimensions and melt much of the Earth's crust or mantle. A projectile the size of Mars would shatter the Earth nearly to its core and eject at least 10 percent of the Earth's mass into surrounding space. If the projectile were as large as the planet Mercury, the Earth would be totally disrupted, leaving little more than the metal core.

Computer models of such catastrophes have focused on the glancing impact of a Mars-sized projectile – that is, a protoplanet with about 10 percent of the mass of Earth. These calculations indicate that the ejected material, coming almost entirely from the Earth's mantle, would be heated to a high temperature. If about one-tenth of the Earth's mass were ejected by an oblique impact, we could expect roughly 10 percent of *this* mass (1 percent of the mass of the Earth) to end up in orbit. This 1 percent could condense and reaggregate to form a satellite – the Moon.

If the Moon formed as the result of such an impact, we can readily see why it is depleted of metal and composed primarily of mantle rock. In addition, the impact heating of this material would drive off any water or other volatiles to produce the conditions we see today. In general, then, the giant impact hypothesis is consistent with the constraints imposed by our detailed knowledge of the Moon and the ways it differs from the Earth; however, much additional work will be needed to confirm the details of the process.

Most planetary scientists think we are converging on the correct answer for the origin of the Moon, but we would be more comfortable with this giant impact theory if we saw other evidence of

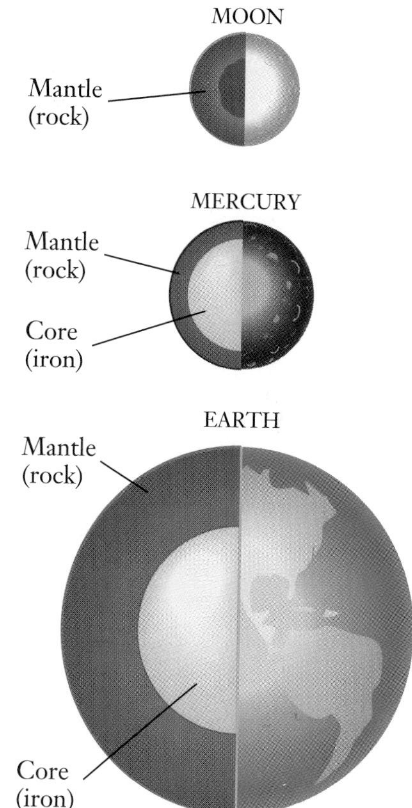

The interior structures of Earth, Moon, and Mercury are quite different. On the Earth, about one-third of the mass is found in the metallic core, with most of the remaining two-thirds in the rocky mantle. The Moon, in contrast, has no metallic core and is composed entirely of rocky material. Mercury is the opposite of the Moon, with a large metal core and thin mantle. In terms of their origins, we can imagine that both the Moon and Mercury are the product of giant impacts, with the Moon representing mantle material ejected from the Earth in a catastrophic event, while Mercury is the remnant of a planet that has lost most of its mantle in one or more large impacts.

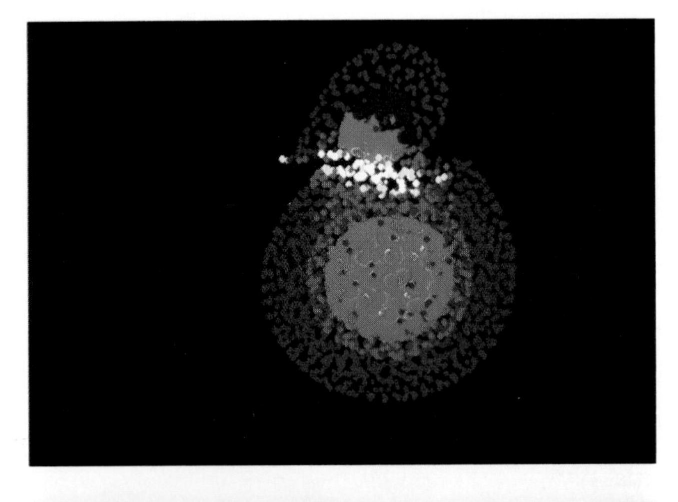

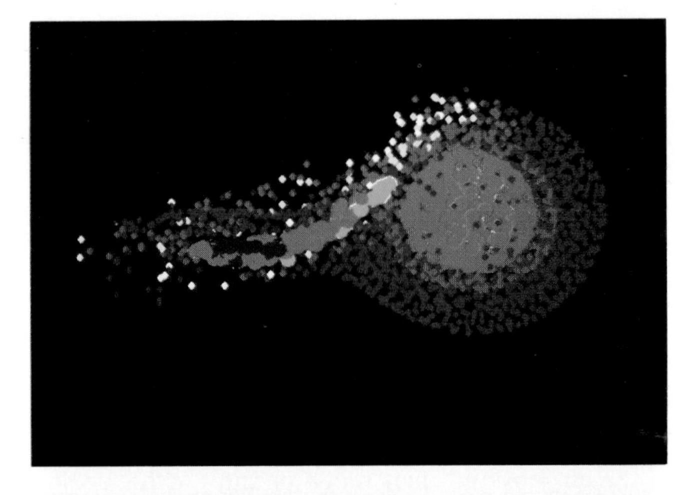

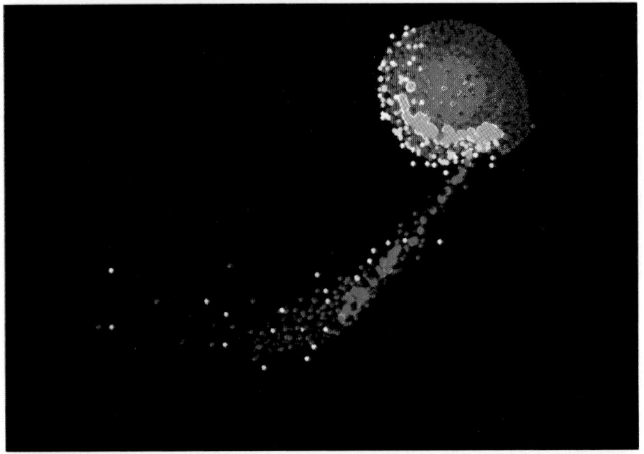

the role of impacts in the early history of the solar system. Consider again the results of impact by a still larger protoplanet on the Earth or other planet of similar size. In such a case it is possible for a planet to lose most of its mantle material yet retain most of its metallic core. This seems like a good recipe for making Mercury. If one giant impact could create a Moon without metal, another, paradoxically, could make a Mercury that is mostly metal. One of the strengths of the giant impact hypothesis is that it offers explanations for the composition of two such different bodies.

The idea of planets and protoplanets whizzing about and colliding with each other is both disturbing and stimulating. While not yet proved, the concepts are gaining increasing acceptance. We should remember, however, that such collisions were confined to the very earliest periods of the solar system. Whatever formed the Moon, we know that our satellite had a solid crust 4.4 billion years ago, within about 100 million years of the origin of the solar system itself. The catastrophic era of planetary history may have been eventful, but it was short-lived.

The Moon was most likely formed from a giant impact with the proto-Earth which took place during the first 100 million years of solar system history. The computer simulation on the facing page shows the consequences of an oblique impact by an object the size of Mars. Blue and green colors indicate iron from the cores of the Earth and the projectile; red, orange, and yellow indicate rocky mantle material. Note that most of the ejected iron falls back onto the Earth, leaving primarily rocky material to coalesce into the Moon. Successive frames in this sequence show increasingly larger volumes of space around the Earth.

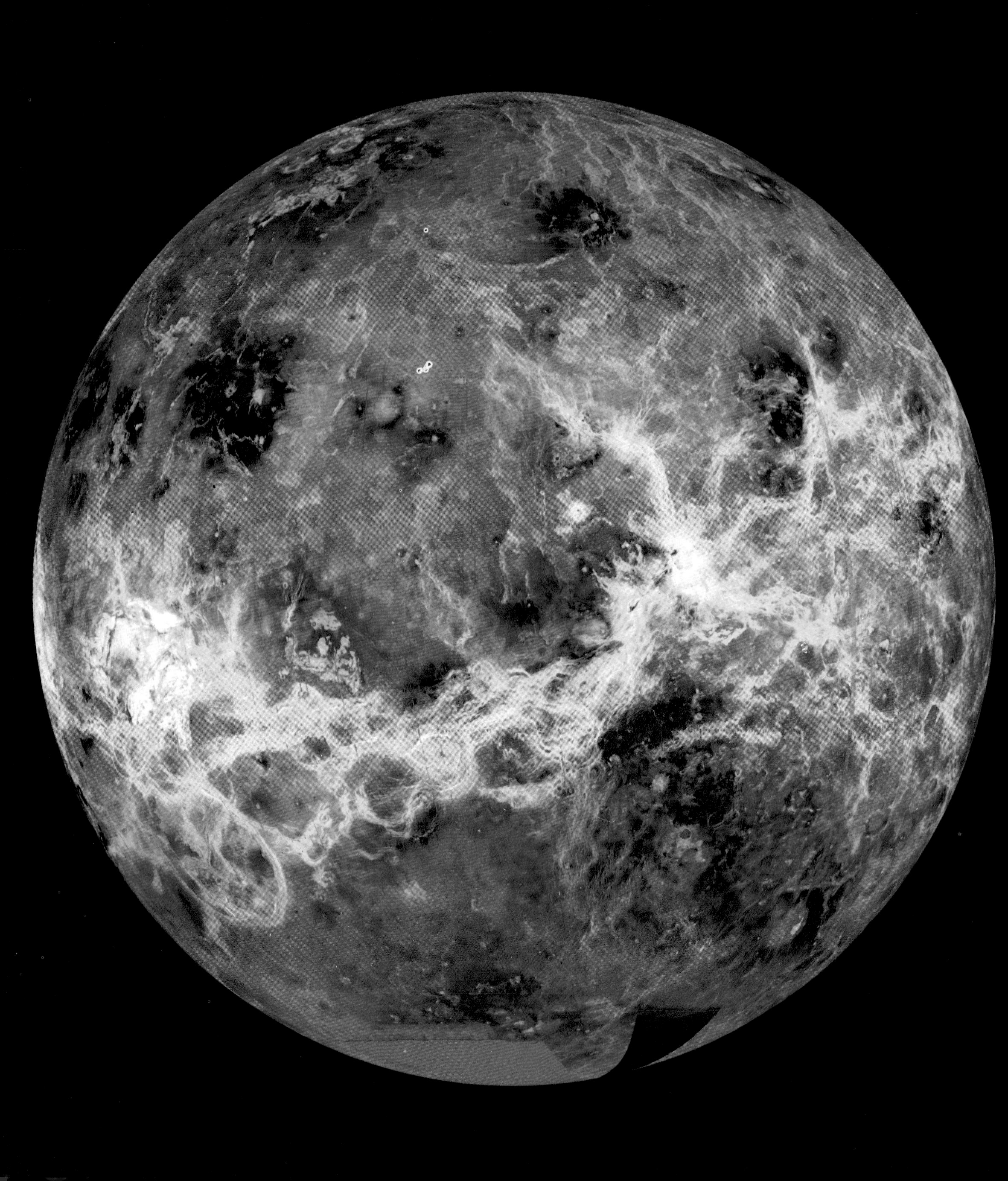

STRANGE TWINS

Venus and Earth

The complex geology of Venus was first revealed by the Magellan radar orbiter, which mapped the entire planet between 1990 and 1992. The photograph on this page shows the spacecraft being released from the Shuttle at the beginning of its voyage to Venus.

No two planets in the solar system are more similar than Venus and Earth. They are celestial twins, only 40 million kilometers apart when they pass each other at 20-month intervals. Both planets have diameters of about 12,000 km and nearly equivalent masses and densities. Since their densities approximate each other, their bulk compositions must be nearly the same. Each has a substantial atmosphere and extensive clouds.

Until fairly recently, astronomers assumed that conditions on the surface of Venus were akin to those on the Earth. Thick clouds in the atmosphere of Venus block our view of the surface, encouraging speculation. Popular astronomy books and even college texts from the middle of this century used illustrations of dinosaurs lolling in verdant swamps as images of possible conditions on Venus. Not until the mid-1950s were these fanciful ideas confronted with the cold — in this case, we should say hot — reality of scientific fact.

The discovery that changed our view was a high level of radio emission from Venus. Every object weakly emits radio static, approximately in proportion to its temperature. Since the atmosphere of Venus is transparent to radio waves, the radio radiation comes directly to us from the surface. When radio telescopes were used to take the temperature of Venus, they revealed a planet with a surface at 450 C — hot enough to melt lead or zinc. This surface temperature was uniform, with neither geographic nor seasonal variations: the entire planet was hotter than the highest noontime temperature on Mercury. Good-bye to the swamps and dinosaurs of Venus!

Additional research during the 1960s revealed that the atmosphere of Venus is huge, with a surface pressure of 90 bars — that is, 90 times the sea-level pressure on Earth. Unlike that of our planet, moreover, this atmosphere is composed mostly of carbon dioxide. This massive CO_2 atmosphere is responsible for the high surface temperature through the greenhouse effect. Application of greenhouse theory to Venus began with the work of a doctoral student named Carl Sagan, working on his thesis at the University of Chicago. In 1959 Sagan showed that if an atmosphere contained a great deal of water vapor, it could trap incoming sunlight, forcing up the surface temperature. Later it was found that the atmosphere of Venus was dry, but Sagan and others used more sophisticated models to demonstrate that a massive CO_2 atmosphere works the same way.

Venus is a most inhospitable planet. Not only is its surface off-limits to humans, but we do not even know how to construct robotic explorers that can withstand its heat and pressure for more than a few hours. Add the corrosive effects of sulfuric acid clouds and the dim red glow in which the surface is bathed and you have a perfect picture of hell. At the surface, Venus is no twin of Earth; it seems at best a distant relative. Yet the overall bulk similarities of the two planets remain. It is to highlight one of the most intriguing questions in planetary science – "How did Venus and Earth come to be so different?" – that we discuss the two planets together in this chapter.

Venus and Earth offer us visibility into an era of solar system history complementary to that discussed in the last chapter. Because of their larger size both planets remain geologically active, and their surfaces are relatively young. The heavy bombardment of the first half-billion years of solar system history, which cratered the lunar highlands, finds no representation in the surviving surfaces of these planets. Indeed, the youngest lunar terrain, the mare lavas erupted 3.2 billion years ago, is older than 99 percent of the rocks on the Earth and, presumably, Venus.

Lifting the Veil of Venus

In spite of its brilliance in the night sky, Venus is a great disappointment to telescopic observers. Featureless clouds frustrate efforts to understand the true nature of the planet beneath. Without radar to penetrate the clouds, we could not even determine the rotation period of the planet. One of the surprises of early radar

Venus is shrouded by clouds at all times. In visible light these clouds are generally featureless, but structure is apparent in reflected ultraviolet light, as shown in this photograph from the Pioneer Venus orbiter. These ultraviolet cloud patterns are related to high wind speeds in the stratosphere of Venus.

work was the discovery that Venus rotates in a retrograde (backward) direction with the remarkably long period of 256 days.

In view of the difficulty of studying Venus from the Earth, it should not surprise us that Venus has been visited by more spacecraft than any other planet, most of them designed to peer beneath its cloudy veil. Exploration began with the U.S. Mariner 2 craft in 1962, which verified the high surface temperatures inferred previously from Earth-based radio observations. A second U.S. flyby took place in 1967. However, the lead in spacecraft exploration of Venus quickly passed to the U.S.S.R., which launched more than a dozen successful missions between 1970 and 1985.

From the beginning, the Russian scientists concentrated on building probes to penetrate the clouds; but because they had underestimated the atmospheric pressure, their early models were all crushed before they could reach the surface. In 1970, however, the Venera 7 probe landed successfully and broadcast for 23 minutes before succumbing to the high temperature. More landings followed, and in 1975 Venera 9 and 10 radioed back the first surface photographs, revealing desolate lava plains. The illumination was about the same as that on Earth under a stormy sky. In 1978, the United States successfully deployed four Pioneer Venus probes into the atmosphere, and one of these survived for 67 minutes after striking the ground.

The Venera and Pioneer Venus probes provided excellent measurements of clouds and atmosphere. In addition, the Venera landers studied surface chemistry and radioed worm's-eye-views of four sites on Venus. None of these missions, however, yielded insight into the global properties of the planet. The first such overview was obtained in the late 1970s by the Pioneer Venus orbiter, which mapped the planet with cloud-piercing radar. Although the radar map had a resolution of only about 100 km, not much better than our naked-eye view of the Moon, it revealed a topography different from that of Earth. Venus was found to consist mostly of plains, with only a few continental masses and no evidence of deep basins like those of the oceans of Earth.

The next step was to use radar imaging to produce pictures of the surface. Venera 15 and 16, which orbited Venus in 1983, carried the first imaging radar and mapped the northern hemisphere with a resolution of 2 km — almost as good as telescopic resolution of the Moon. In 1984, the Russians achieved another first with a pair of instrumented balloons deployed in the atmosphere. This was to be the final Soviet mission, however. Satisfied with their many successes at Venus, they turned their attention outward toward Mars and new challenges.

One more mission to Venus must be reported, however. Launched by the United States in 1989, Magellan carried an

The hellish surface of Venus has been photographed from several landers in the Russian Venera series. This wide-angle panorama from Venera 13 shows a rocky plain with the spacecraft itself visible at lower center. The temperature is hot enough to melt lead, and the scene is suffused by a reddish glow of sunlight penetrating through the thick overcast.

Highlights in the Exploration of Venus

Date	Spacecraft	Mission	Accomplishments
1962	Mariner 2	flyby	first successful interplanetary mission
1967	Venera 4	probe	first data from Venus atmosphere
1972	Venera 8	lander	first data on surface composition
1975	Venera 9	lander	first photographs from surface
1978	Pioneer 12	orbiter	first radar map
1978	Pioneer 13	probes (4)	most detailed atmospheric analysis
1983	Venera 15	orbiter	first radar imaging from orbit (2-km resolution)
1984	VEGA 1	lander	first instrumented balloons in atmosphere
1990	Magellan	orbiter	most detailed radar imaging (100-m resolution)

advanced radar imaging system with a resolution of 100 m, some 20 times better than the radar on Venera 15 and 16. It began mapping Venus in September 1990 from a polar orbit, acquiring one 20-km-wide strip from pole to pole each day and building up the map gradually as the planet rotated beneath. By the summer of 1992, it had imaged more than 98 percent of the surface – an area equal to the combined land areas of Earth, Mars, Mercury, and the Moon. The total data returned by Magellan exceeded that of all other planetary spacecraft together.

Thanks to Magellan, our knowledge of the surface topography of Venus is better even than of the Earth. There are places on our planet, beneath the oceans or ice caps, where we do not have equivalent maps at 100-m resolution. Of course, in many other ways we know more about our own planet, but in the following discussion, we will try not to be overwhelmed by the wealth of information we have about the Earth. We will not describe most contemporary processes – the water cycle, the carbon cycle, or recurrent episodes of glaciation – in order to adopt a longer-term perspective and to compare Earth and Venus at similar levels of detail.

The surface of Venus has been revealed by imaging radar that can penetrate beneath the clouds. This is a computer-generated oblique radar view of the planet reconstructed from Magellan data. The two large volcanoes are Sapas Mons in the center and Maat Mons on the horizon. In this representation the vertical relief is exaggerated by 10 times and a false color has been added to the image.

Global Comparison

Venus and Earth are each terrestrial planets composed primarily of the elements silicon, oxygen, iron, magnesium, aluminum, nickel, and sulfur. Their chemistry is oxidizing, with very little hydrogen. Each planet was heated early in its history, perhaps as it accreted from the original solar nebula. Heating led to differentiation, the dense iron, nickel, and much of the sulfur sinking to form a core, while various silicate rocks rose to form a mantle and crust. By definition, the mantle is the part that remains hot and somewhat plastic, while the crust is a solid layer of low-density rock that floats on top of the mantle.

We can determine the interior structure of the Earth by studying the behavior of seismic waves generated by earthquakes, which reveal the density and temperature at various depths and allow us to infer chemical composition. We find that the metal core of the Earth accounts for about one-third of the mass of the planet; the mantle constitutes most of the remaining two-thirds, with the crust amounting to less than 1 percent of the total. At least part of the core is liquid, and turbulent motions within this spinning core generate the Earth's magnetic field. In contrast, Venus has no measurable magnetic field, perhaps because its core lacks turbulent motions. The central parts of both planets still retain a part of their primordial heat, while the decay of radioactive elements in their mantles provides a continuing source of energy to power geological activity.

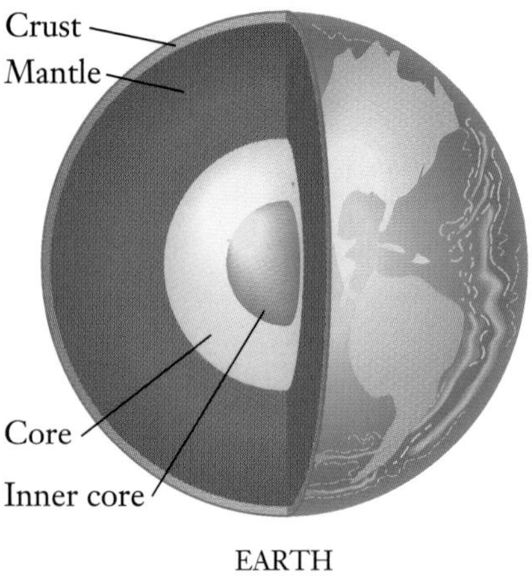

Crust
Mantle
Core
Inner core

EARTH

This drawing illustrates the interior structure of the Earth, but it could just as easily be Venus, based on our current understanding of the two planets. Both objects differentiated early in their history to form a metallic core (primarily iron and nickel) that accounts for about one-third of the mass of the planet. On Earth there is a solid inner core and a liquid outer core. The bulk of the planet consists of silicate rocks in the mantle.

When we look at a map of the Earth, we are immediately struck by the division into ocean and land. Even without the presence of water, a distinction would remain. Approximately two-fifths of the Earth's surface area consists of continental highlands (measured to the edges of the continental shelves), while three-fifths consist of deep basins.

The distinction between the Earth's continents and ocean basins extends to the composition and thickness of the crust. The continents are typically made of granite, a type of igneous silicate rock formed from the slow subsurface cooling of lava. This solid granitic layer is from 50 to 70 km thick. Parts of the continental masses form folded mountain ranges thrust as high as 10 km above sea level. In contrast,

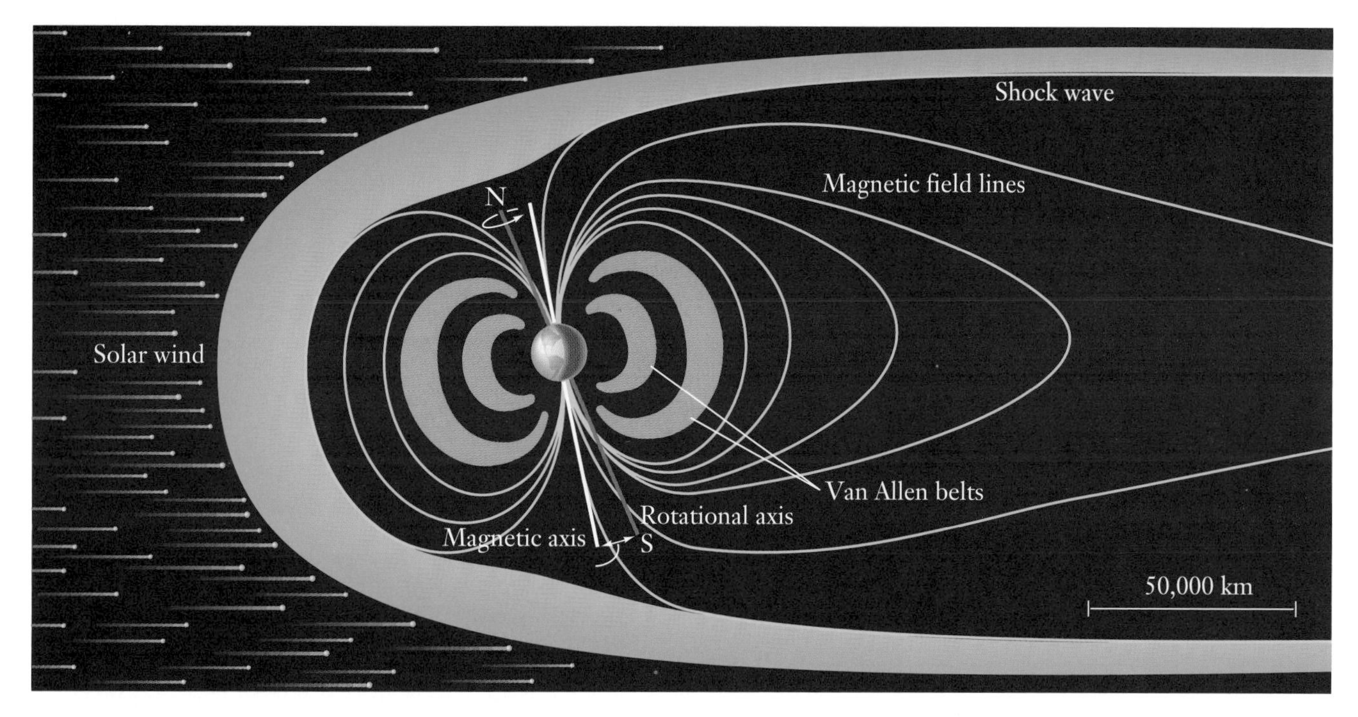

The Earth has a substantial magnetic field generated in its metallic core, and this magnetic field in turn supports a magnetosphere extending outward 10 to 20 times the radius of the planet. Energetic ions trapped in the inner part of this magnetosphere constitute the Van Allen Belts, which were discovered in 1958 from data sent back by the first U.S. satellite, Explorer 1. Sometimes these are called radiation belts, but the term is misleading; the belts are composed of charged particles, not radiation.

the oceanic crust is only about 6 km thick and is composed of basalt, similar to the basaltic lavas of the lunar maria. Most of the Earth's volcanic activity, moreover, is concentrated in the ocean basins, where more than a cubic kilometer of new lava is extruded every year. Constant volcanic renewal of the oceans makes them among the youngest features of the planet, with typical ages of about 100 million years – about 30 times less than the ages of the basaltic maria on the Moon. The deepest parts of the basins are about 10 km below sea level, 20 km lower than the highest continental mountains.

The surface topography of Venus reveals a different picture. More than five-sixths of the planet consists of gently rolling volcanic plains with a total vertical scale of only about 2 km. There are no extensive basins comparable to the ocean basins of the Earth. Higher regions, mostly aggregated into two large continents, Aphrodite and Ishtar, occupy the remaining one-sixth of the surface. Aphrodite, an equatorial continent the size of Africa, stretches nearly one-third of the way around the planet. Ishtar is smaller, about the size of

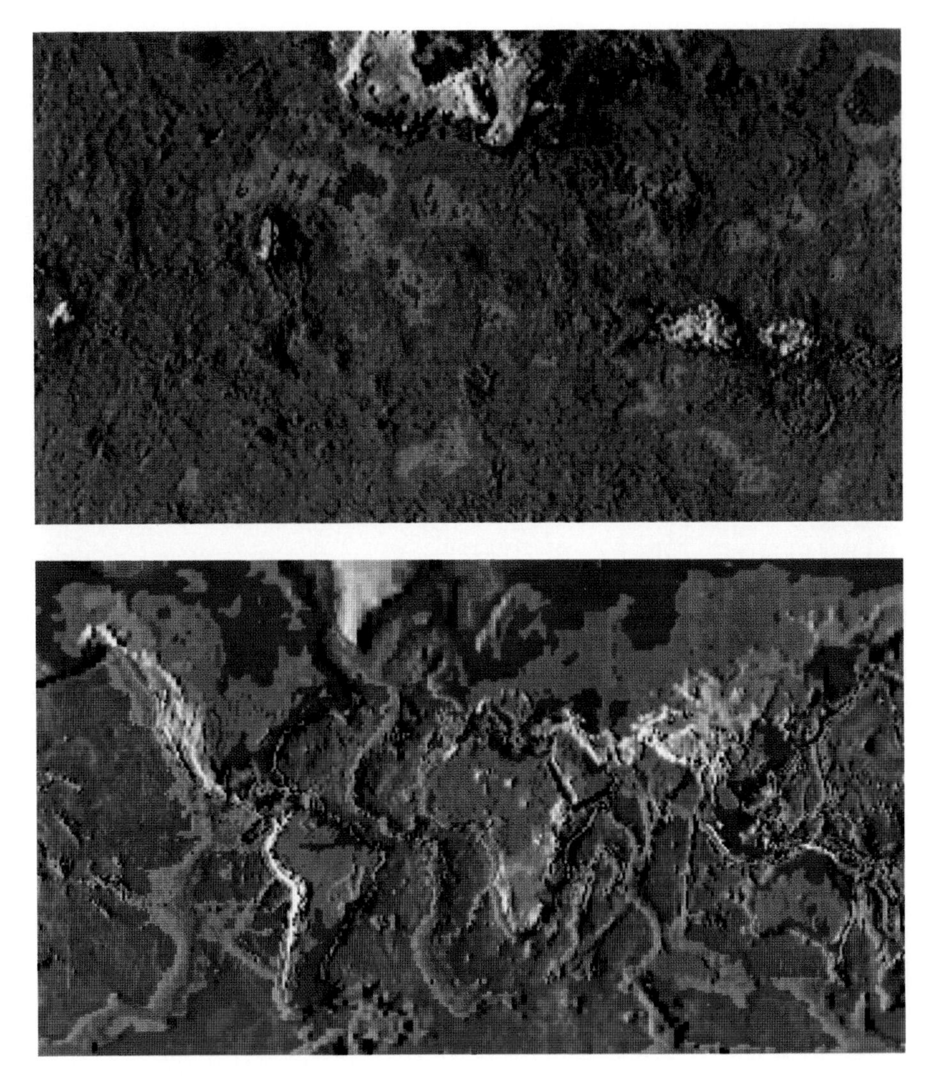

The large-scale topography of Venus is very different from that of the Earth. As shown in this Pioneer Venus radar map (top, resolution about 50 km), most of the planet consists of rolling lowlands, with just two large continents: Ishtar in the north and Aphrodite near the equator. The size of Ishtar has been exaggerated in this mercator projection. The bottom image is of the Earth at the same scale, clearly showing the division of our planet into continental highlands and deep ocean basins. In both images, the highest topographic points are white.

Australia, and is located at the latitude of northern Europe on the Earth.

Aphrodite, part of a band of folded and fractured crust that extends around much of the equator of Venus, is not unlike the continents of the Earth in general appearance. Ishtar, which is higher, resembles the Himalayan Plateau of the Earth and includes the highest mountains of Venus, rising to an altitude of 11 km, roughly the height of Mt. Everest above sea level on Earth.

One of the unexpected results of the Magellan mission is the discovery that there is a planetwide change in surface chemistry associated with altitude. Essentially all regions higher than 5 km have exceptionally high radar reflectivity. Looking at the maps you might think these highland regions were covered with snow, but in this case we are looking at the ability to reflect 12-cm radar waves, not visible light. One suggested explanation is that the iron-bearing mineral magnetite, present at lower altitudes, is chemically converted to an electrically conducting mineral called pyrrhotite at the cooler temperatures above 5 km altitude. Because it is conducting, pyrrhotite exhibits the high radar reflectivity observed by Magellan.

Craters and Volcanoes

As we saw in discussing the Moon and Mercury, the presence of impact craters provides a measure of surface age. Terrestrial erosion and sedimentation erase craters almost as fast as they are formed. Only a handful can be identified easily on our planet, although nearly 150 ancient, degraded craters have been recognized from space imagery. The sparsity of craters testifies to the high level of geologic activity on our restless planet.

The Magellan survey of Venus reveals about 900 impact craters, most of them larger than 20 km in diameter. Projectiles that would make craters much smaller than this do not survive passage through the dense atmosphere of Venus. From the numbers of craters we can infer that the average age of the surface is about 500 million years—youthful in comparison with the lunar maria, but substantially older than the ocean basins on Earth.

The largest known impact craters on the Earth are Chicxulub in Mexico, with a diameter of 180 km, and the more complex Vredefort structure in South Africa, which may be the result of multiple impacts. The largest crater on Venus is Mead, with a diameter of 280 km. These are about the same size as the largest lunar post-mare craters. There is no indication on any planet of craters larger than 300 km formed during the past 3 billion years.

One of the most striking aspects of the craters on Venus is their pristine appearance. Almost all look as if they were formed just

The presence of many large, fresh impact craters on Venus is a consequence of the older surface and lower erosion rates on that planet, compared with the Earth. This Magellan radar image shows three craters in the Lavinia region of Venus, the largest with a diameter of 50 km. The rough crater rims and ejecta are excellent radar reflectors and therefore appear bright in a radar image.

yesterday, geologically speaking, with fewer than 5 percent distorted or degraded by volcanic activity or other geologic processes. These crisp craters prove that there is little erosion or sedimentation on Venus. Evidently, the processes that ultimately destroy craters on Venus do not act gradually; rather, the craters are wiped away in a short time, presumably by large-scale volcanism. The surface can

Cone-shaped volcanoes are more common on Earth than on Venus. This is an aerial view of Karymskii Volcano in Kamchatka (Russia), an active volcano about 1.5 km high. This type of volcano is constructed from the fallback of incandescent lava (pyroclastics) ejected under pressure from the vent.

survive almost unaltered for hundreds of millions of years, only to be suddenly flooded by lava — perhaps by a single massive eruption.

Much of the volcanic activity on Venus takes the form of basaltic eruptions that inundate large areas, much as the mare volcanism flooded the impact basins on the near side of the Moon. The Earth also experiences such eruptions from time to time; probably the most dramatic terrestrial example is the eruption about 65 million years ago that produced the Deccan Plateau, which covers much of the southern half of India. A more recent case is provided by the Snake River Plateau of eastern Washington, whose thick, mare-like lavas can be seen in striking cross-section along the gorge of the Columbia River.

More familiar to most of us are the eruptions that produce volcanic mountains. For many, the word volcano conjures up an image of a steep, cone-shaped mountain with a crater at the summit — Vesuvius in Italy, Fuji in Japan, or Pinatubo in the Philippines. These volcanoes are the product of fire-fountains of fine-grained lava that shoot up from the vent propelled by hot gas. Sometimes the eruptions are more violent, and a large cloud of so-called volcanic ash is expelled. Falling debris from such eruptions construct these steep-walled mountains or smaller cinder cones.

A second major type of volcano results from the gentler eruption of highly fluid lavas that contain less dissolved gas. Repeated surges of free-flowing lava gradually build up broad, shallow-sloped mountains, called shield volcanoes after their resemblance to a medieval knight's shield laid on its side. Shield volcanos are among the largest landforms on the Earth – one pair, Mauna Kea and Mauna Loa in Hawaii, together form a mountain more than 200 km across rising 9 km above the ocean floor. Mauna Loa is still growing; eruptions take place almost continuously from its summit and flanks.

Volcanoes are as common on Venus as on the Earth. The largest are shield volcanoes; about 150 greater than 100 km in diameter have been identified. The biggest shields include Sif, Gula, and Maat, comparable in width to Mauna Loa and Mauna Kea, but only about half as high. In one of the largest areas of shield-type volcanism, Beta Regio, there are indications of currently active eruptions. Few steep-

Broad shield volcanoes are common on Venus. This radar image shows the shield volcanoes named Sif and Gula, each about 4 km high. Surrounding this volcanic complex are many tectonic cracks that appear as bright streaks in the radar images.

sided volcanic cones exist on Venus, indicating that the more explosive types of eruptions are suppressed by the high pressure of the massive atmosphere.

Several results of volcanism on Venus differ from anything on the Earth. One of these has been called pancake domes. Several dozen of these distinctive features have been found, each consisting of an almost perfectly circular, flat dome with steep sides. Typical widths are 25 km, with heights of 2 km. Each pancake dome appears to be made of highly viscous lava erupted rather suddenly from a single vent, like a giant belch from the mantle of Venus (domes of viscous lava also form on Earth, but they are much smaller and far less symmetrical). At the opposite extreme are lava rivers of extremely low viscosity. About 40 of these channels are longer than 100 km, and one, called Hildr, is 7000 km long—as long as the longest rivers on Earth, such as the Nile or the Mississippi. By contrast, the longest lava channels on Earth extend only a few tens of kilometers.

The most distinctive volcanic features of Venus are the circular structures called coronae. Several hundred are known, with diameters

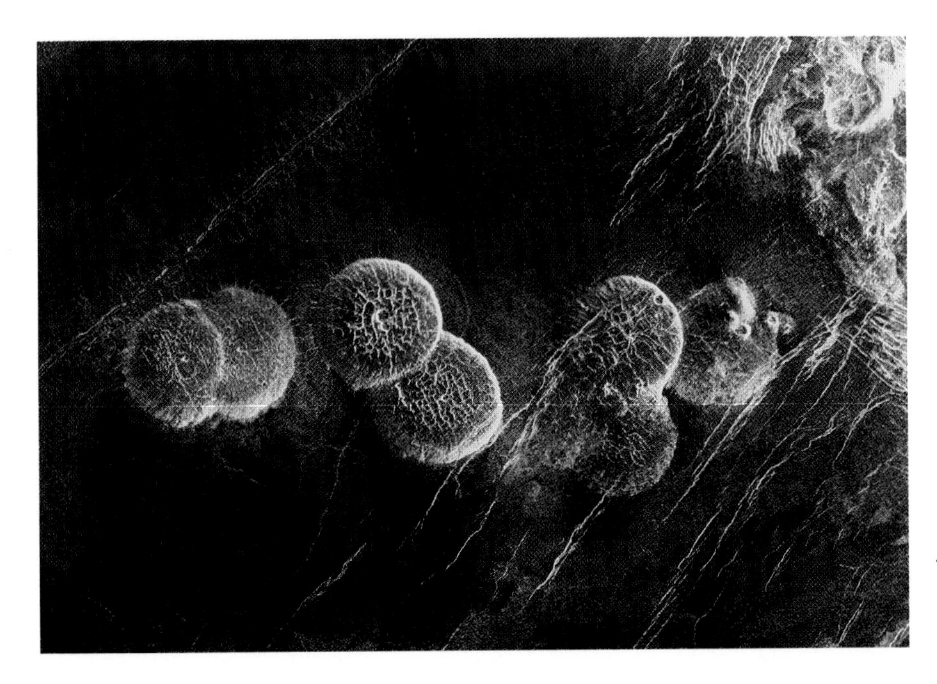

These unusual volcanoes are called pancake domes. Each dome is about 25 km in diameter and about 2 km high. Such circular features are thought to be formed of highly viscous lava, erupted within a short span of time. The Earth also has domes of viscous lava, but they are smaller and less symmetric.

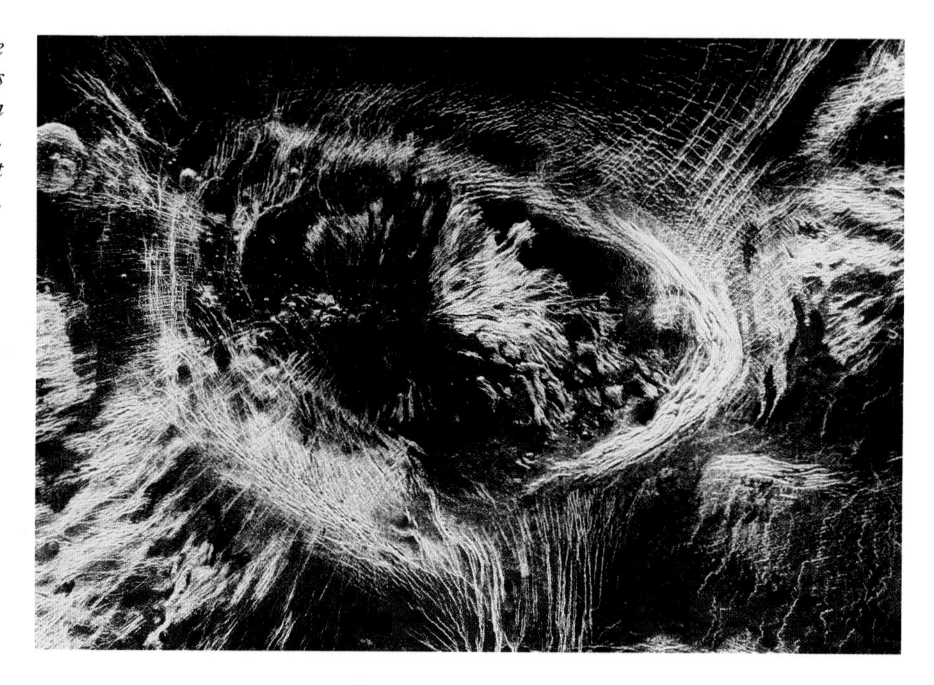

On Venus, rising plumes of mantle material often fail to burst forth as surface volcanoes. Instead they form complex circular features called coronae. This Magellan radar image is of Bahet Corona, which is about 1000 km across.

up to 2000 km for Artemis, the largest. A corona typically has a slightly raised interior surrounded by a low circular ridge and a trough, or moat. Each appears to be the result of a mantle plume that became inactive before it could form a true shield volcano — an example of stillborn volcanism that can be spotted easily on Venus because of the absence of obscuring erosion.

Plate Tectonics

The origin of the large-scale landforms on Earth and Venus is one of the most basic issues in the geology of the two planets. We see a superficial resemblance: both planets have continental highlands, volcanoes, a few high mountains, and a scattering of impact craters. Venus, however, lacks the ocean basins that are an important feature of terrestrial geology, its continents are smaller than those of the Earth, and it displays the distinctive coronae. How did these features originate, and what determines the similarities and differences between the two planets?

Pick up an old geology book and you will find accurate descriptions of the Earth but little perspective on the processes that mold the terrestrial surface. Prior to the middle of the twentieth century, scientists did not understand the fundamental forces that produce terrestrial ocean basins, continents, or mountain ranges. There were references to rising and falling sea levels, episodes of mountain building, and dramatic changes in climate (rain forests once flourished in the Antarctic), but no common pattern to lend coherence to these ideas. Now, however, we make sense out of Earth history by using the concept of plate tectonics.

Tectonics is the name geologists give to stresses acting in the crust of the Earth – stresses that can squeeze together or pull apart the crustal rocks, often accompanied by earthquakes and volcanic eruptions. It has been clear for two centuries that such stresses can produce mountains where the crust is compressed or valleys where it is extended. The revolution in twentieth-century geology, however, resulted from the perception that the crust of the Earth is composed of a dozen or so large sections (plates) like the pieces of a jigsaw

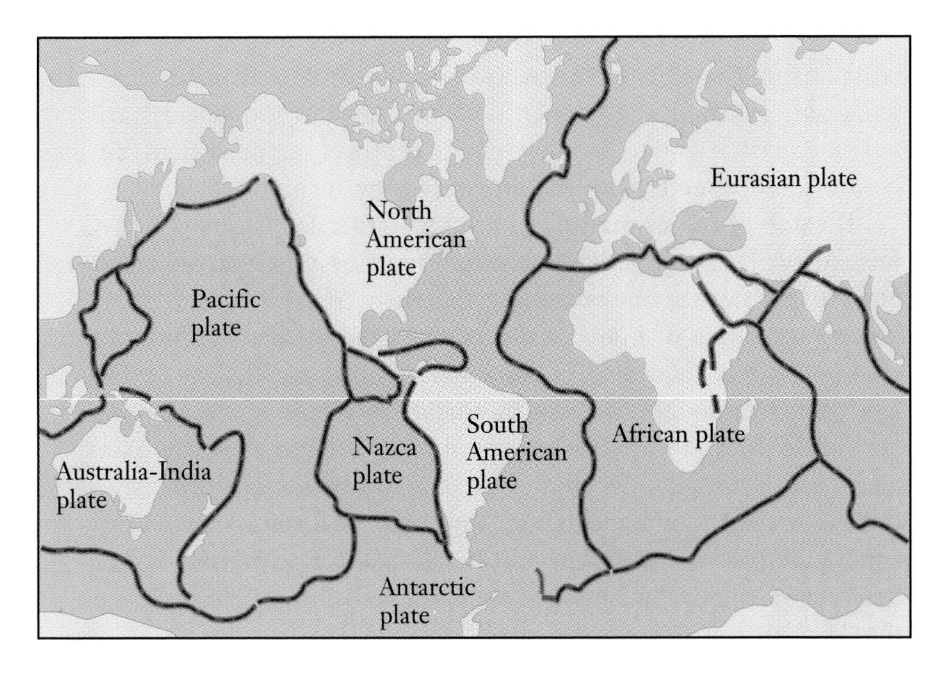

Terrestrial geology is dominated by plate tectonics. The crust of the planet consists of about a dozen large blocks, or plates, that move about on the surface, propelled by slow forces of convection in the underlying mantle.

puzzle, each of which, driven by slow currents in the underlying mantle, can move with respect to the others.

Plate tectonics had its origin at the beginning of the century when Alfred Wegener, a German astronomer and meteorologist, suggested that the geologic similarities between the coasts of North America and Europe, and between the coasts of South America and Africa, could be understood if the continents of the New World had once been connected to those of the Old. He proposed that the Atlantic Ocean was a relatively young feature, formed in the widening gap between the two hemispheres. Wegener called his theory continental drift, and he extended it to explain other similarities between continental masses now widely separated. The theory was rejected by most geologists, however, because its proponents could suggest no plausible force to explain the migration of continents across the solid oceanic crust, to which they seemed firmly anchored.

Continental drift became accepted and was incorporated into the broader theory of plate tectonics when, in the 1960s, observations clearly showed that the Atlantic Ocean was widening as a line of volcanoes, approximately midway between the two continental masses, injected new lava along the ocean floor. Scientists realized that the continents do not drift over an unyielding crust, but rather that the plates carrying the continents are being forced apart by new crust arising between them. (A location where crustal plates are separating is called a rift zone.) The two sides of the Atlantic are moving apart at about five meters per century—a speed that can now be measured directly with satellite surveying techniques.

If some of the crustal plates are separating, or rifting, by the formation of new oceanic crust, there must be other places on the planet where plates are crunching together. At such boundaries, plates can collide or one can slip beneath the other. A collision, most common where each plate is thick (continental crust), lifts up the land to make mountains. One spectacular example is the Himalayas, generated where the Indian subcontinent bumps against the Asian plate. But if one plate is composed of thin oceanic crust, it may slip under the plate against which it is pushing to form a subduction zone; examples are found around much of the Pacific rim. Subduction regions are marked by earthquakes and volcanism, hence the "ring of fire" circling the Pacific from Alaska to the Philippines.

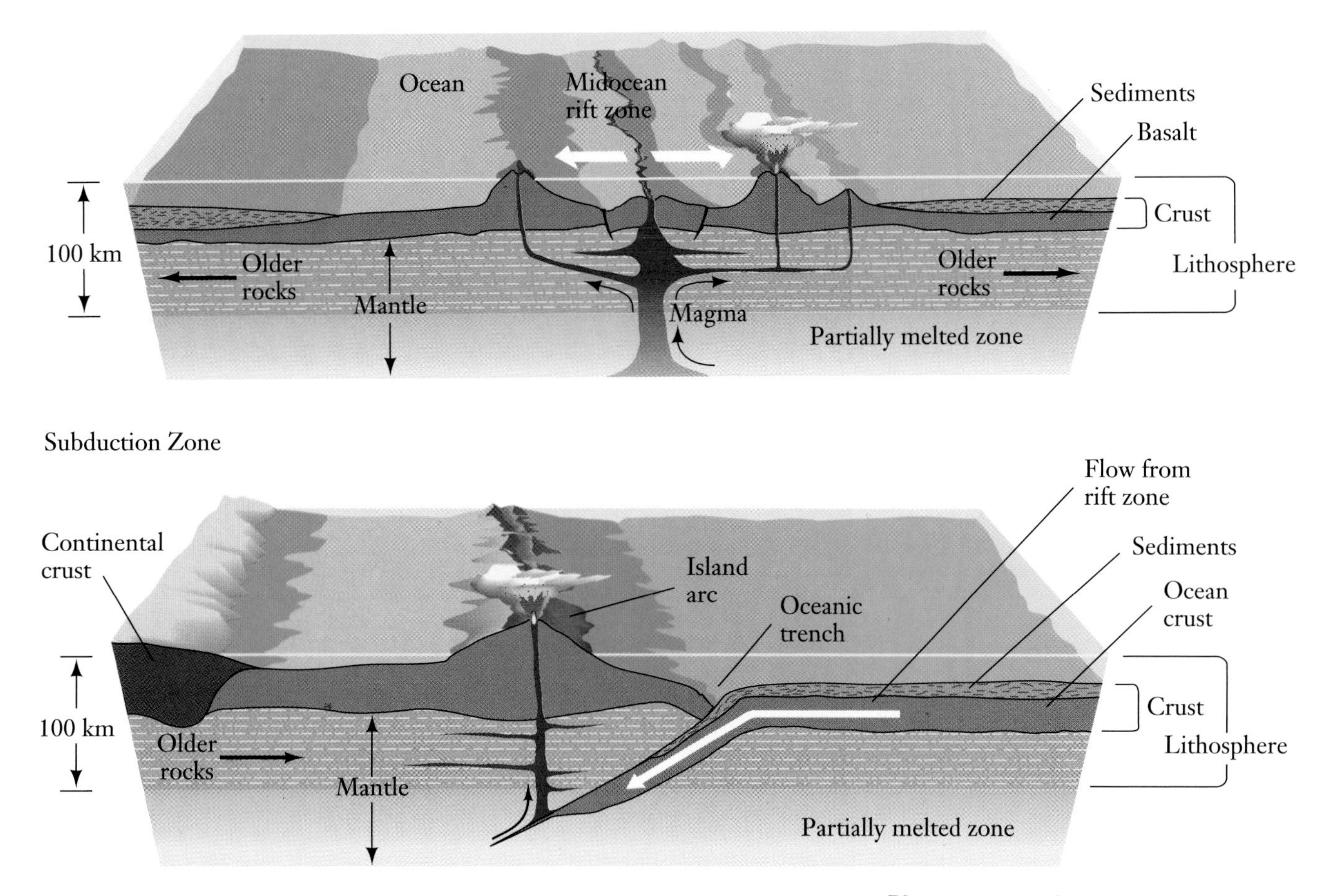

Plate tectonics is driven by convection currents in the mantle, which inject volcanic magma at rift zones and force plates apart. There are about 60,000 km of active rifts on Earth, most in ocean basins. When a thin oceanic plate is driven into a continental plate it is often subducted, passing beneath the continental plate and forming a deep-ocean trench at the boundary. The subducted oceanic sediment is heated at great depth, releasing CO_2 and driving volcanic eruptions to produce an offshore island arc.

At some intersections, the plates neither separate nor converge, but slide along relative to one another. The result is a fault zone along the plate boundary. Since plates do not slip easily past each other, fault zones tend to be the source of earthquakes. A well-known example is the San Andreas Fault in California, marking the boundary between the North American and Pacific plates along a line from Baja California to San Francisco. These two plates move five to six meters per century relative to one another. At the town of

The most spectacular mountains on Earth are sculpted into fantastic shapes by the erosive forces of water and ice, neither of which is present on other planets. These remarkable peaks are part of Cerro Torre in Patagonia, Argentina.

Parkfield on the San Andreas, the result is a moderate earthquake about every 22 years, each accompanied by about a meter of slippage. In contrast, there has not been major movement along the fault near Los Angeles since the great earthquake of 1848; when that part of the fault finally gives way, the lateral motion could be as much as 7 m, sufficient to generate a large earthquake. Along plate boundaries, the longer the interval between earthquakes, the more destructive they are likely to be.

Most major terrestrial landforms owe their existence directly or indirectly to the internal forces that drive the motion of crustal plates. Heat escaping from the interior rises in large, semistationary plumes beneath rift zones or other volcanic hot spots. In compensation, cooler material descends into the mantle beneath subduction zones. The scale of these convection currents is comparable to the dimensions of the plates themselves. This fact, together with the strength and internal stability of the plates, maintains the slow motions that shape continents, form mountains, and generate the volcanism that continually renews the ocean floors.

Tectonics on Venus

We might expect that Venus, similar in size and internal structure, would experience plate tectonics like the Earth's. The Magellan radar maps permit us to test this hypothesis for the first time.

The level of volcanism on Venus — about the same as on the Earth — together with the broad geographic distribution of volcanoes on Venus suggests escape of heat from the interior, probably driving some kind of mantle convection. The coronae of Venus further signal the presence of mantle hot spots. Coronae and shield volcanoes are surface expressions of plumes of rising material. In the case of the coronae, the plume top is centered under the central rise, while material seems to be descending under the surrounding moat to form minisubduction zones. However, the scale of these convection plumes on Venus appears to be smaller than the thousands of kilometers that characterize the crustal plates on the Earth. To distinguish it from terrestrial plate tectonics, this phenomenon has been called blob tectonics.

Virtually the entire crust of Venus is subject to tectonic forces. Everywhere we look we see ridges and cracks, readily visible in the absence of overlying soil or vegetation. From the rolling lowlands to the large shield volcanoes, Venus shows evidence that its crust is being pushed and pulled by interior forces.

Tectonic activity is not distributed evenly over the crust, however, but concentrated in a broad tectonic belt that stretches around the

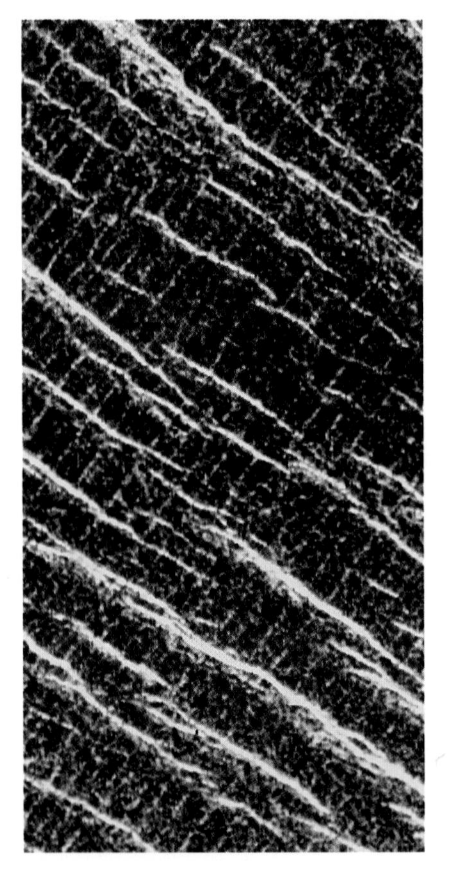

Because there is little soil and no vegetation to confuse us, we can see the tectonic patterns on Venus much more easily than on our own planet. This region of the Lakshmi Plains has been fractured to produce a striking grid of cracks and ridges separated by about a kilometer.

equator and includes the Aphrodite continent and a prominent region of low mountains called Alpha Regio. Within this belt the crust has been compressed and folded to create a dense network of ridges and mountains. Although the altitudes are not as great, this tectonic belt looks similar to many continental regions of Earth, produced by the pressure of one crustal plate on another. There are also a few — but very few — trenches that resemble terrestrial subduction zones. Similarly, a few small rift zones exist in regions of volcanic activity, but nothing approaching the midocean ridges of Earth. One has the impression of a superposition of numerous small-scale centers of tectonic activity, as though the mantle convection currents are smaller or the thick crust lacks the mobility of the tectonic plates on Earth.

The region of Venus that most closely resembles terrestrial mountains is the Ishtar continent, which includes the Maxwell Mountains (the highest range on Venus) and the broad Lakshmi Plateau. The similarity of these features to the Himalayan Mountains and Tibetan Plateau strongly suggests that they are the product of tectonic compression of the crust. Measurements of the steep mountains bordering Lakshmi, which can exceed 30 degrees in slope, demonstrates that these mountains are receiving active support from below, since left to themselves they would slump under their own weight. Compression seems to be the simplest way of achieving such support. However, there is a very different possibility, namely that Ishtar sits atop a particularly large and active mantle plume. Geologists are currently debating the merits of these two alternatives.

In summary, Venus is similar to the Earth in its high levels of volcanic and tectonic activity. Plumes of hot material rise through the mantle and stress the crust to form folded mountains, faults, subduction regions, and a wide variety of volcanic landforms. Yet Venus has not developed the planetary-scale plate tectonics that lead to the large continents and deep ocean basins we have on Earth.

Atmospheres

While the geology of Venus is generally similar to that of Earth, the atmosphere is very different, although we expect that the two planets began with similar atmospheres about 4 billion years ago. Their

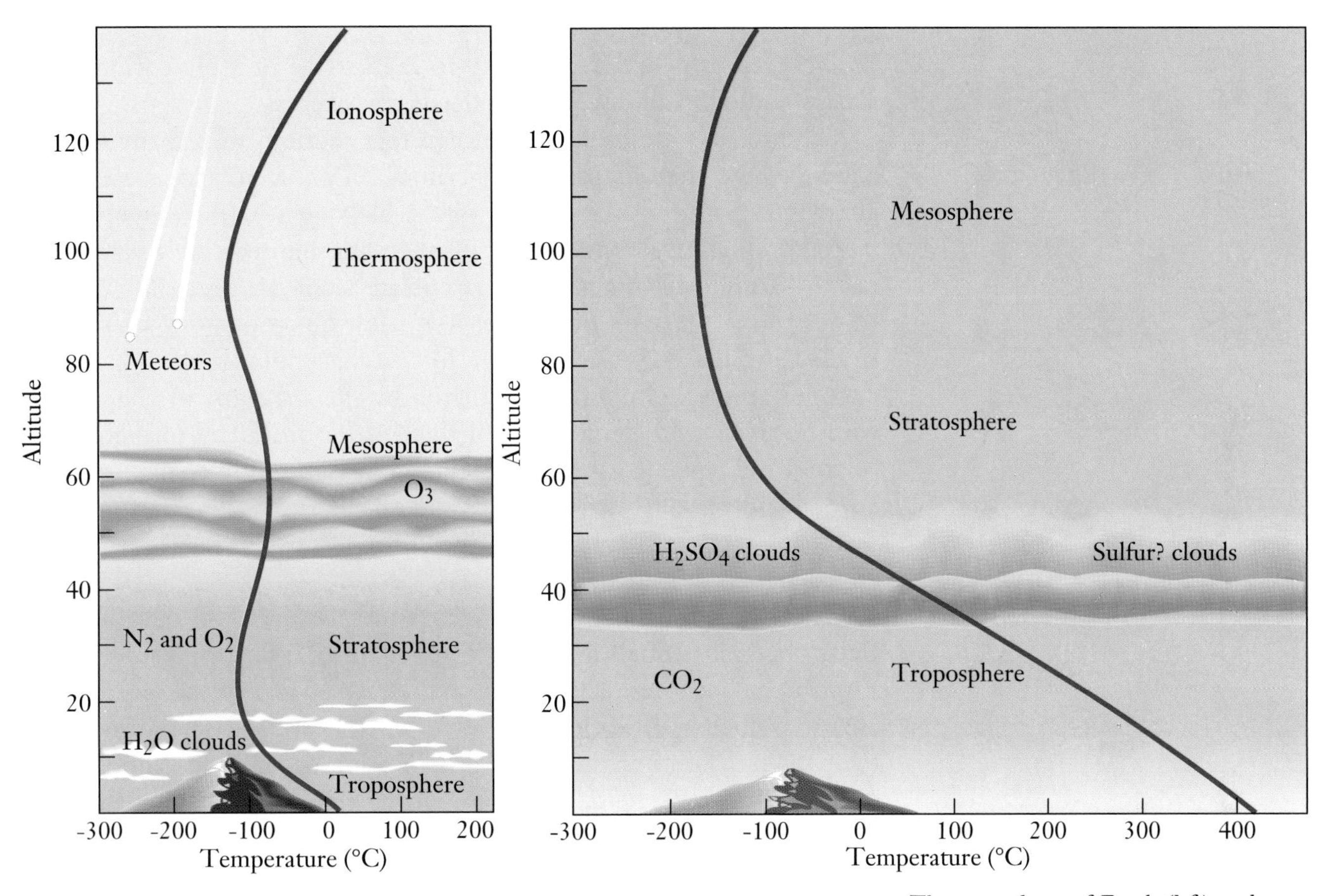

The atmospheres of Earth (left) and Venus (right) are very different, in spite of other similarities between the two planets. The massive CO_2 atmosphere on Venus produces a much stronger greenhouse effect than on our planet, with correspondingly elevated surface temperature. Only in their stratospheres (at altitudes of about 30 km on Earth and 80 km on Venus) do the two atmospheres have similar temperatures and pressures. The thick clouds of Venus are composed in part of H_2SO_4, which may be emitted by volcanic eruptions.

divergent atmospheric evolution is intimately coupled with the emergence of life on Earth; however, it is difficult to determine whether a suitable atmosphere encouraged the development of life or whether the presence of life controlled the evolution of the atmosphere. Probably both statements are correct.

Each planet has a substantial envelope of gas that extends more than 100 km above the surface. On Earth, the atmosphere is partially transparent, although at any time about half the planet is obscured by clouds. These clouds are mostly confined to the turbulent lower part of the atmosphere, the troposphere, which is about 10 km thick. On Venus, the troposphere extends up to about 60 km altitude and supports an unbroken layer of clouds between about 40 and 60 km.

Sunlight never penetrates directly to the surface of Venus, which is always bathed in diffuse light under heavy overcast.

The Earth's atmosphere is in constant motion, with many small-scale disturbances and storms superimposed on such large-scale circulation patterns as the trade winds, blowing steadily at subtropical latitudes. The resulting weather is inherently unpredictable, as we know. While we can generally extrapolate from the present to determine the state of the weather a few hours or even a few days in the future, accurate long-range forecasts are impossible. This complex atmospheric motion is driven by the transport of energy from warmer to cooler regions of the surface: from day to night, or from the equator toward the poles. The constantly changing diurnal and seasonal deposition of sunlight is the engine that maintains the winds and powers the storms of the terrestrial atmosphere.

Circulation in the atmosphere of Venus is similar to that of Earth, but substantially simpler. The length of a day on Venus is 116 Earth days, and there are no seasonal effects. Most of the solar energy is absorbed by clouds at about 50 km altitude, so the surface temperature does not vary from one location to another. The large mass of the atmosphere, nearly a hundred times greater than the Earth's, further smooths out turbulence and encourages stability. All these effects join to maintain a steady atmospheric circulation pattern almost devoid of "weather." Surface winds are very low, resulting in little erosion, as we have noted from the radar images. Only at the top of the troposphere are there strong winds, with a planet-girdling jet stream that blows at about 150 m/s, three times faster than the jet streams on Earth.

Primary constituents of the Earth's atmosphere are nitrogen (78 percent) and oxygen (21 percent). As a stable and chemically nonreactive element, nitrogen is the sort of gas we might expect in a planetary atmosphere. Oxygen, on the other hand, highly reactive and unstable, would not be present if not for its constant replenishment by photosynthesis. Two significant minor constituents are carbon dioxide (0.03 percent but growing) and water vapor (variable in quantity but important because it can condense to produce clouds). The atmosphere exerts a surface pressure of about 10 tons per square meter: by definition, 1 bar of pressure.

The atmosphere of our planet is constantly in motion, driven by the unequal deposition of sunlight and amplified by the exchange of water between the atmosphere and surface. This photograph shows a large cyclonic storm as seen from orbit. Such tropical storms, known variously as hurricanes or typhoons, can generate high surface winds and intense rainfall.

Venus's much more massive atmosphere, with a surface pressure of 90 bars, is composed primarily of carbon dioxide (96 percent) and nitrogen (3 percent). There is no oxygen worth mentioning, and very little water vapor. Because of the high surface temperature, there can be no liquid water on the surface either, making Venus a remarkably dry planet. In the absence of water vapor, Venus has no water clouds and no rain. Instead, its clouds are composed of tiny particles of sulfuric acid and, perhaps, elemental sulfur. Sulfuric acid discharged into Earth's atmosphere by volcanoes is quickly washed out by rain, sparing us the corrosive acid clouds of Venus.

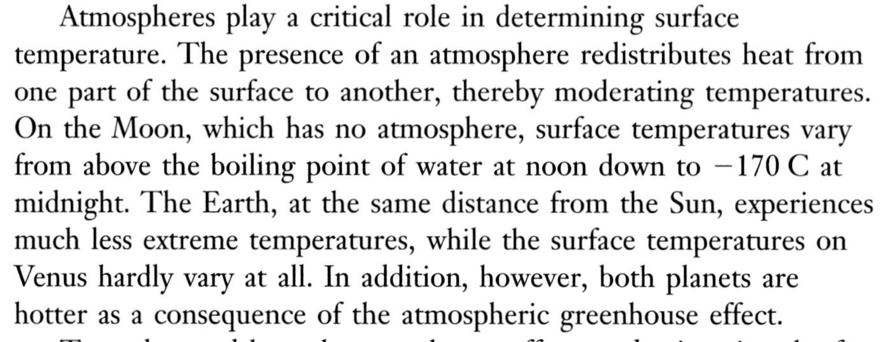

Atmospheres play a critical role in determining surface temperature. The presence of an atmosphere redistributes heat from one part of the surface to another, thereby moderating temperatures. On the Moon, which has no atmosphere, surface temperatures vary from above the boiling point of water at noon down to −170 C at midnight. The Earth, at the same distance from the Sun, experiences much less extreme temperatures, while the surface temperatures on Venus hardly vary at all. In addition, however, both planets are hotter as a consequence of the atmospheric greenhouse effect.

To understand how the greenhouse effect works, imagine the fate of sunlight that diffuses through the atmosphere of Earth or Venus and is absorbed at the surface. The surface temperature represents an equilibrium between this incoming solar energy and the energy that the surface can send back into space in the form of infrared or heat radiation. Carbon dioxide and other so-called greenhouse gases are opaque to this infrared radiation and act as a blanket, holding in the heat. As a result, the surface temperature rises until it reaches a new equilibrium between incoming and outgoing energy. The greater the infrared opacity of the atmosphere, the more the surface temperature is enhanced.

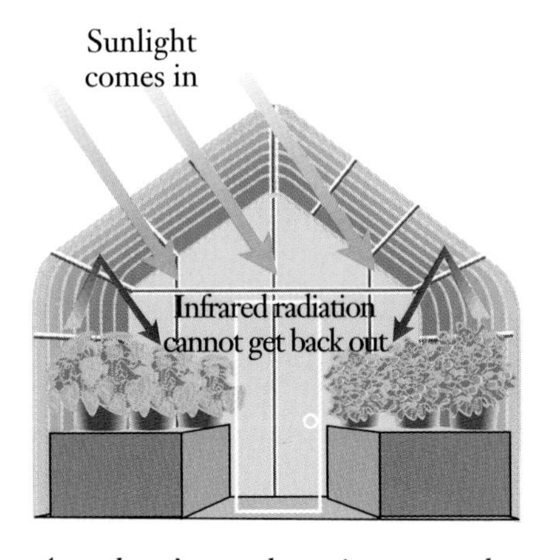

A gardener's greenhouse is warmer than the outside air partly because the glass in its roof transmits visible sunlight but is opaque to infrared radiation emitted by the interior. The infrared radiation is trapped inside, raising the temperature. The same effect takes place in the atmospheres of Earth and Venus, but with the infrared opacity provided primarily by CO_2 gas.

The greenhouse effect received its name because a gardener's greenhouse is heated in a similar way, except that the infrared opacity is provided by the glass walls and roof of the greenhouse rather than by atmospheric CO_2. Glass is transparent to visible light but blocks infrared, trapping energy inside. If you have access to a car with a closed trunk, you can test the greenhouse effect for yourself by leaving it in sunlight with the windows up. Then compare the temperature in the passenger compartment, where the greenhouse effect has been in action, with the trunk, where it has not. You may be surprised at how much cooler the trunk is.

On Earth, the greenhouse effect is responsible for our average surface temperature remaining above the freezing point of water; without it, the global average would be about 20 C lower and most of the oceans would freeze. The effect on Venus, much greater because of its massive CO_2 atmosphere, results in a surface temperature of 450 C – about 400 C higher than it would be in the absence of the atmosphere. The large CO_2 atmosphere, therefore, dictates the hellish surface conditions on our sister planet.

Atmospheric Evolution

In addition to the sizes of their atmospheres, the Earth and Venus display three major chemical differences that need to be explained. The Earth is depleted of carbon dioxide but has excessive oxygen, while Venus is depleted of water. Assuming that they began with similar composition, how did this situation develop? We cannot provide definitive answers to this question, but the scenario described in this section is probably correct in outline, although almost certainly deficient in detail. In any case it provides us with a plausible framework for comparing the chemical history of the two planets.

Our current understanding of the origin of the terrestrial planets suggests that the building blocks from which they formed were primarily rocky and metallic planetesimals containing little water or carbon dioxide. Some water was chemically bound into the minerals of these planetesimals, but the quantity was insufficient to account for the oceans of the Earth. Subsequently, impacts of comets from the outer solar system brought in additional water, carbon dioxide, carbon monoxide, and traces of methane and ammonia to form the bulk of the atmospheres and oceans for Venus, Earth, and Mars.

Initially these atmospheres were dominated by carbon dioxide, with lesser quantities of carbon monoxide and nitrogen; if surface temperatures were below the boiling point of water, oceans must have accumulated. As far as we can tell, similar situations should have obtained on Venus, Earth, and Mars during the first few hundred million years of solar system history.

Being closest to the Sun, Venus was the warmest. The atmospheric greenhouse effect raised temperatures and boiled away the oceans, if they were able to condense in the first place. The result was a massive and unstable atmosphere of water vapor, most of which quickly escaped from the planet. The remaining water was broken down by sunlight into hydrogen and oxygen, the hydrogen escaping and the oxygen combining with carbon monoxide to form additional carbon dioxide until Venus reached its current dry, hot state.

On Earth, lower initial temperatures permitted water to remain liquid and reduced the strength of the greenhouse effect. Conditions favored surface chemical reactions that reduced the amount of atmospheric CO_2; as a consequence, a moderate atmosphere developed, containing nitrogen, carbon monoxide, and trace constituents of such reducing gases as ammonia and methane. Organic compounds brought in by comet impacts survived to accumulate in the warm seas. Sometime around 4 billion years ago these conditions led to the development of life.

Initially life was confined to the molecular level. These early self-replicating molecules probably obtained their energy from the rich marine mixtures of organic compounds or from inorganic sources like the submarine hot springs recently discovered in the deep oceans. Eventually, however, increasing biological sophistication led to the development of single-celled creatures, some of which were able to use the energy of sunlight to manufacture new organic material. By 3 billion years ago these photosynthetic bacteria were building large colonies called stromatolites. Stromatolites, one of the most successful life forms, still thrive in warm, shallow coastal seas.

The proliferation of photosynthetic life began the process of atmospheric evolution that has led to the world we know. Release of oxygen eliminated any remaining atmospheric carbon monoxide or reducing compounds to create a more oxidizing environment. By 2 billion years ago, with some free oxygen present in the atmosphere, animals could develop with metabolic activity based on oxygen. More efficient chemical pathways led to increased diversity and larger, multicelled creatures. Eventually sufficient atmospheric oxygen accumulated to shield the surface from lethal ultraviolet radiation with a protective ozone layer.

Ozone, a form of oxygen with three atoms per molecule rather than the usual two, is formed by the action of sunlight on oxygen at altitudes of about 30 km. Although ozone is a minor constituent of the atmosphere, it is critical to the biological history of our planet. In the absence of ozone, life would be confined to the seas. By absorbing essentially all of the solar ultraviolet light, ozone made possible the colonization of the land and air. Beginning less than a billion years ago, life could begin to emerge onto the heretofore barren rocks of the continents. Thus life gave our world oxygen, and oxygen made the world safe for life.

Meanwhile life was also affecting carbon dioxide, the primary constituent of the early atmosphere. Many tiny marine creatures evolved a capability to extract CO_2 from the water and use it to manufacture protective shells of carbonate. When these creatures died, their shells sank, amassing into thick layers of carbonate sediment on the ocean floors. The result was a gradual depletion of

Stromatolites are among the most successful of the early life forms on our planet. These colonies of microorganisms have flourished in warm, shallow seas for nearly 3 billion years. These stromatolites are from the Great Bahama Bank.

Much of the ocean floor is covered by a calcareous ooze consisting primarily of the shells of marine microorganisms. This photo shows microscopic foraminifera collected from the ocean bottom. Such deposits, slowly converted to limestone, provide a storehouse for a great deal of terrestrial carbon that would otherwise be in the atmosphere in the form of carbon dioxide.

atmospheric CO_2 as available carbon was chemically bound into the crustal rocks. Eventually CO_2 became a trace constituent of the atmosphere, now dominated by nitrogen and oxygen.

Estimates of the quantity of carbonates in ocean sediments suggest that this carbon, if returned to the atmosphere, would produce enough CO_2 to exert an atmospheric pressure of 70 bars. In the absence of life, the Earth would have about the same balance of CO_2 and nitrogen as does Venus, with a correspondingly enhanced greenhouse effect. The presence of oxygen, the near absence of CO_2, and the moderate greenhouse effect and surface temperatures on our planet can all be traced to the influence of life on the evolution of the atmosphere.

Uniqueness of Earth

Earth is the only planet with liquid water on its surface and oxygen in its atmosphere. It is also the only planet with living organisms. As we have seen, these unique aspects of our world are coupled. Earth also appears to be the only planet with plate tectonics, and there is speculation that this phenomenon is also related to the nature of our atmosphere and oceans. Some scientists believe that the ability of the crustal plates to slide as they do over the mantle is the result of lubrication by water and the large temperature gradient across the Earth's crust; neither condition prevails on Venus.

As noted above, one of the important consequences of life has been the removal of carbon dioxide from the atmosphere – most in the form of marine sediments, but a smaller fraction buried as fossil fuels: coal, peat, oil, and natural gas. It is interesting to note that the steady reduction in atmospheric carbon dioxide has coincided with a slow increase in the luminosity of the Sun as it evolves. Thus the change over time in the solar energy striking Earth has been compensated by a steadily decreasing greenhouse effect, so that the surface temperature of the planet has remained almost constant for the past 4 billion years.

Unfortunately, because modern industrial society depends on energy extracted from fossil fuels, we seem determined to accelerate the recycling of fossil carbon, extracted over hundreds of millions of years, back into the atmosphere. The amount of CO_2 in the atmosphere currently increases by nearly 1 percent per year. The inevitable result will be an enhanced greenhouse effect, but no one knows the magnitude of this enhancement. Will temperatures increase by 2 degrees, or 5, or even 10? Fundamental issues of public

Human consumption of fossil fuels and deforestation in South America and Africa are having important consequences for the atmosphere and climate of the Earth. The most easily measured effect is an increase in atmospheric CO_2, as documented in measurements made for the past 35 years at an observatory high on the slopes of Mauna Loa in Hawaii. In addition to an annual cycle, we can see that the total quantity of the gas increased by 10 percent between 1958 and 1988. There is some evidence for global warming during the twentieth century as well, especially since 1980, but it is difficult to determine to what extent this is the result of enhanced atmospheric CO_2.

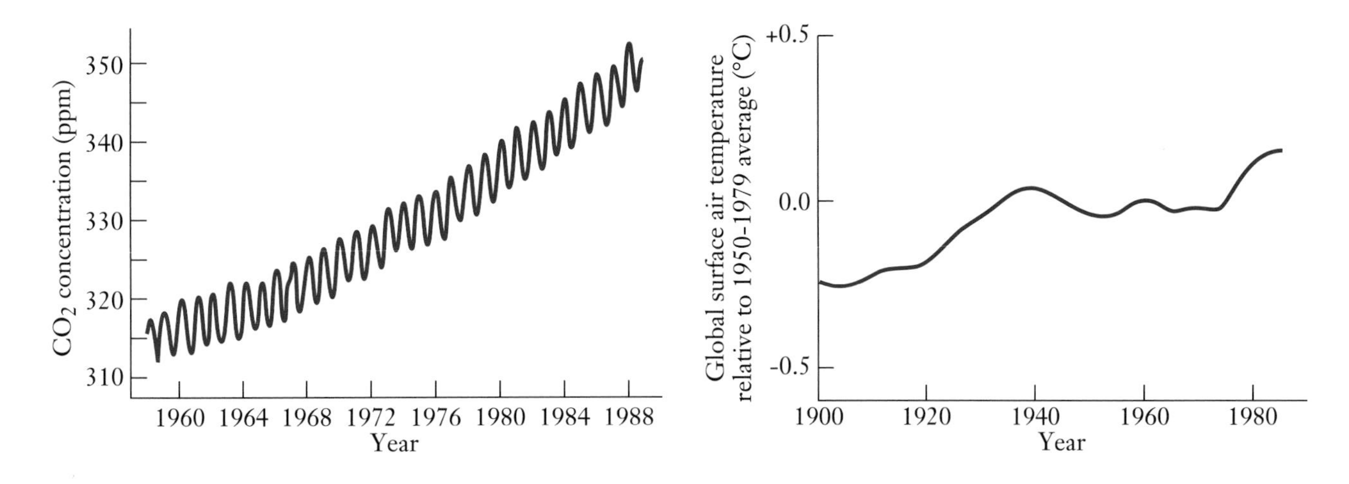

policy depend on the answers to such questions. Some participants in the debate seem to question the reality of the greenhouse effect, but this is foolish. If not for the greenhouse effect, the Earth would be frozen in a deep global ice age. The question is not the reality of the effect, but the degree to which it is being enhanced by release of more CO_2 into the atmosphere.

Study of the atmosphere of Venus — an extreme example of the greenhouse effect at work — may help us to ask whether the Earth could suffer the same fate. Imagine increasing the temperature of the Earth to the point where some of the carbon in the crust is released into the atmosphere. The more carbon is released, the hotter it gets; and the hotter it gets, the more carbon is released. Eventually temperatures could rise above the boiling point of water, and the oceans would become part of the atmosphere, contributing further to the warming. We can see, in principle, how such a runaway greenhouse effect might turn paradise into hell, transforming the Earth into a facsimile of Venus. Could this really happen if global warming is pushed too hard? Probably not, but no one knows for sure.

Industrial society is also tampering with the ozone layer. Manufactured chemicals, especially the refrigerants called chlorofluorocarbons (CFCs), have a remarkable ability to destroy ozone in the upper atmosphere; already nearly 10 percent of the ozone is gone at some seasons. We are working strenuously to phase out the manufacture of CFCs, but the CFCs in the atmosphere will still persist a century from now. It is difficult to predict the consequences of increased ultraviolet light for life on Earth, but the effects are sure to be harmful to some degree.

Until human activity began to affect the balance of nature, the Earth had evolved in ways that apparently favor the continued development and proliferation of life. Today life forms are to be found in virtually every environment of the planet, from boiling hot springs to the cold interiors of rocks in the Antarctic.

While many of the large-scale landforms on Venus have structures similar to those on Earth, their detailed appearance is quite different. The Earth is covered with soil and vegetation, neither of which is present on Venus. Terrestrial mountains have sharp peaks

Human disregard for the planet reached a crescendo in the aftermath of the 1991 Persian Gulf War, as hundreds of unchecked oil-well fires in Kuwait consumed fossil carbon to generate CO_2 *and millions of tons of soot.*

and deep valleys that cannot form on Venus, where ice and water erosion are absent. Two-thirds of the surface of our planet is covered with water, but Venus is dry. Earth experiences rain and rainbows, blue sky and bright sunlight—all denied to Venus by its dry atmosphere and thick clouds. From our planet we can look out at night to see the stars and speculate upon our place in the universe, while the inhabitants of Venus, if there were any, would never know that a universe existed beyond their planet.

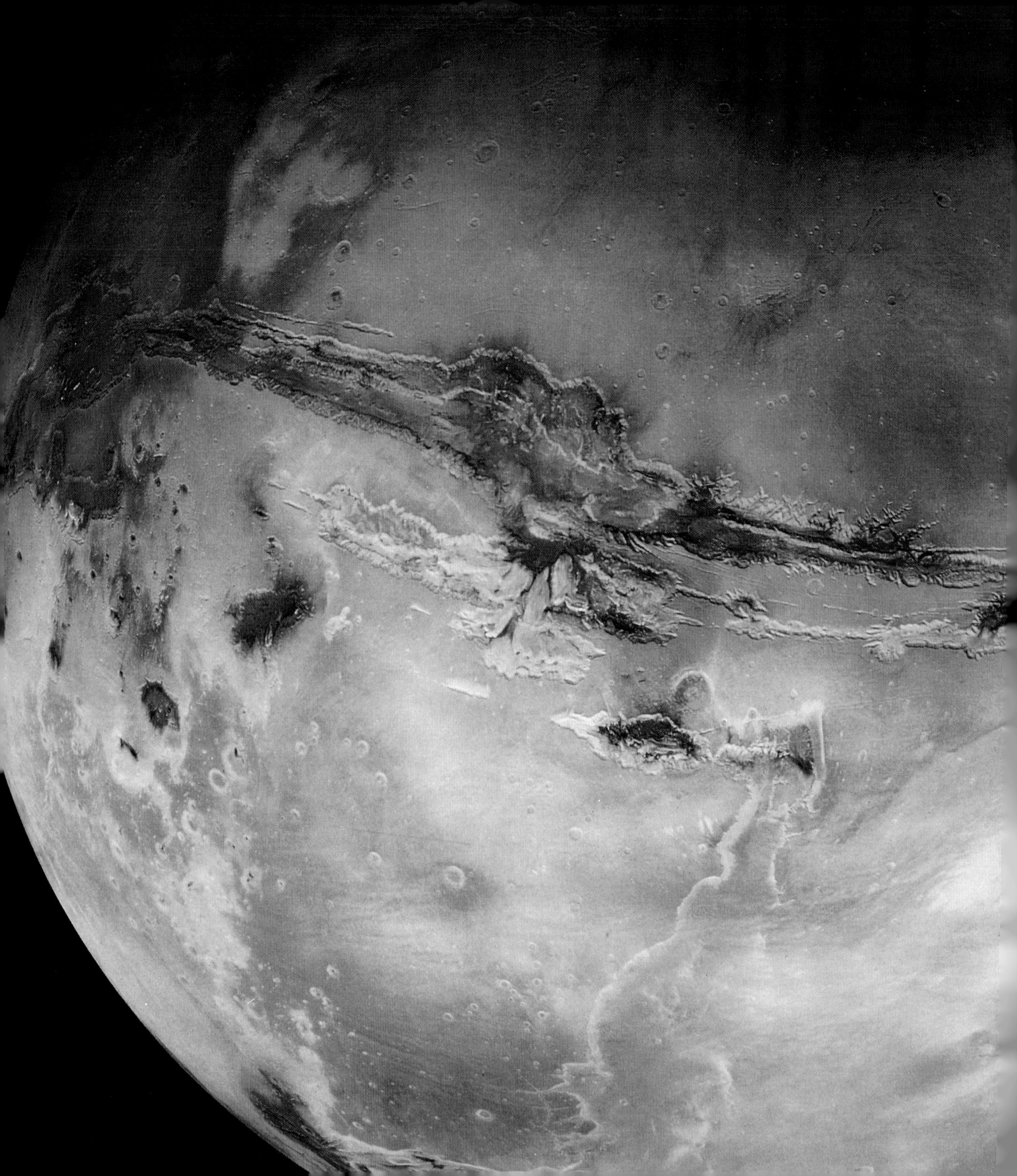

DESTINATION MARS

The Planet Most Like Earth

•••

Although cold and desiccated today, Mars remains the most hospitable of all of the planets beyond Earth. The global view on the facing page was assembled from hundreds of photographs taken by the Viking orbiters; shown here is one of the Viking landers that explored the surface of Mars in 1976.

The "red planet," harbinger of war, a dying world, home to grotesque alien creatures – how many images does Mars invoke? Where does this world fit in the pattern of our deepest yearnings and fears? A century ago, the Victorian science writer and novelist Herbert George Wells made an indelible impression upon public consciousness with his vivid images of a "War of the Worlds" – invasion of Earth by invincible fighting machines controlled by intelligent Martians. Half a century later the Russian-born psychiatrist Immanuel Velikovsky captured popular interest with his fantasies of colliding worlds, attributing the destruction of Troy and the decline of Greek civilization to an errant Mars, thrust from its proper orbit. Today, a cult of pseudoscientists publicizes a mile-long "face on Mars" and interprets it as the artifact of an ancient interplanetary civilization. Supermarket tabloids tell us that Mars is the base for fleets of UFOs, and they even claim that the planet boasts a statue of Elvis! At the same time, the president of the United States promulgates an official policy to send humans to Mars, and proposals for cooperation in Mars missions are discussed at international summit meetings.

Mars plays a major role in both science fantasy and science fact. Its orbit is closer to the Earth than any other planet except Venus, and it is the only planet in which we can recognize a pale reflection of our own world. Like the Earth, it has both a solid surface and an atmosphere, alternately transparent and cloudy. Its day is only half an hour longer than our own, and with a polar axis tilted at 25 degrees, Mars experiences seasons very like the Earth's. White polar caps form during the winter and recede with the coming of spring. The red color of Mars is reminiscent of the orange and ocher sands of terrestrial deserts, especially as we see them in photographs of our planet taken by orbiting astronauts. Even before the era of spacecraft exploration, it was clear that if any environment existed in the solar system that humans might find tolerable, it would surely be on Mars.

Much of the public interest in Mars was stimulated by a remarkable astronomer and publicist, Percival Lowell, who popularized the idea of martian canals. Lowell did not invent the canals, which were first reported in 1877 by the Italian observer Giovanni Schiaparelli as faint dark lines glimpsed near the limit of telescopic detectability. But Lowell took up the subject, founding an observatory in 1894 for the study of Mars and bringing his ideas to the public through books and lectures. Lowell was convinced that the canals proved the existence of intelligent Martians, struggling to preserve their civilization in the face of ecological catastrophe. No matter that most astronomers could not see the canals and no one ever succeeded in photographing them — the image of a species facing extinction was powerful, and the idea that humanity shared the solar

Bostonian Percival Lowell (1855–1916) popularized the idea of canals on Mars. Early in the present century, his work attracted widespread attention to planetary studies, but later — after canals were discredited — his influence on the field was perceived as largely negative.

Lowell depicted Mars as covered with a network of fine geometric lines. Because they were straight, he interpreted these canals as the work of Martian engineers. Now we realize the canals had no objective reality, beyond the tendency of the human mind to seek (and find) order in nature.

system with another intelligent race exerted an influence that extended far beyond the technical controversy over the canals themselves. Perhaps the ghost of Percival Lowell played a part when NASA designated the search for life on Mars as the primary objective of the Viking lander mission of 1976.

Today Mars is cold and inhospitable, seared by lethal ultraviolet light and swept by global dust storms. Yet this planet was not always so uninviting. Long ago it supported a more clement climate, where rain fell and rivers flowed. Discovering that Mars once had a more Earth-like environment ranks as a major accomplishment of the space age. The hope that we may someday find life there, extant or extinct, provides one of the primary challenges to the continuing exploration of our solar system.

Mars holds up a mirror to the Earth. Like Venus, it exemplifies a possible end point of planetary evolution — the frozen desert planet, locked in a terminal ice age.

Exploring the Red Planet

Seen through a telescope, Mars is actually a small orangish disk smudged with a few faint markings and perhaps a white polar cap. Unlike Venus, it has a generally transparent atmosphere, but occasionally the dark smudges and even the polar caps disappear as great dust storms obscure the surface. A thermocouple placed at the focus of a telescope reveals that the surface temperature is below the freezing point of water even at midday. The patient observer will see changes from hour to hour as the planet rotates, and from week to week as the seasons progress; but that is all. Mars is too far away to reveal even a hint of its geology or surface conditions to the telescopic observer.

The finest terrestrial telescope can separate or resolve details as small as about $\frac{1}{4}$ arcsecond, or $\frac{1}{15,000}$ of a degree. When Mars is closest to the Earth, at a distance of approximately 60 million km, this tiny angle corresponds to about 70 km, or 1 percent of the apparent diameter of the planet — a resolution similar to that of the early Pioneer Venus radar map or the naked-eye view of our Moon. Since most geologic features such as mountains, valleys, or impact craters are smaller than 70 km across, they cannot be seen at the resolution corresponding to telescopic images of Mars. To see more, one must approach the planet more closely.

The first interplanetary spacecraft equipped with a camera was Mariner 4, launched toward Mars in 1964. Its primary objective was to photograph the planet at close range in order to reveal its geologic character. In those days, spacecraft cameras were relatively primitive, and data transmission over interplanetary distance was limited. As the Mariner spacecraft flew past Mars it sent back about a dozen photographs covering less than 1 percent of the martian surface. This small sample was enough, however, to change fundamentally our thinking about the red planet, for the only recognizable features in these pictures were impact craters.

The obvious interpretation was that Mars was geologically inactive, with an old surface like that of the Moon. The mission scientists reported that Mars seemed to be "geologically dead." The media, looking for brief but dramatic headlines, reported simply that Mars was a "dead planet." There were no canals, and apparently no life. Further, the spacecraft had measured the atmosphere of Mars and found it to be less than 1 percent as dense as that of the Earth.

Highlights in the Exploration of Mars

Date	Spacecraft	Mission	Accomplishments
1962	Mars 1	flyby	first spacecraft to Mars; unsuccessful
1964	Mariner 4	flyby	first surface photographs; atmospheric analysis
1969	Mariner 6	flyby	high-resolution photographs; polar-cap composition
1971	Mariner 9	orbiter	first planetary orbiter; global survey
1976	Viking 1	orbiter	high-resolution orbital survey
1976	Viking 1	lander	first lander; search for life on Mars
1989	Phobos 2	orbiter	Phobos emphasis; only partially successful
1993	Mars Observer	orbiter	atmosphere and climate emphasis
1995(?)	Mars 94	lander	to deploy instrumented balloon, small landers

Disappointment was widespread, especially among those whose thinking about Mars was still influenced by Percival Lowell.

Additional Mariner spacecraft flew past Mars in 1969, obtaining improved data but not changing the fundamental conclusions of Mariner 4. To achieve a major breakthrough, it was necessary to orbit the planet and carry out a comprehensive study, a goal achieved by the Mariner 9 spacecraft in 1971.

If Mariner 9 had been a flyby like the previous missions, it would have been a complete failure, for it arrived when the planet was shrouded in one of its periodic dust storms. A few weeks after arrival, however, the sky began to clear and the spacecraft cameras could see the ground, although faintly, like the view of a smoggy city from a high-flying aircraft. Eventually the dust settled, and Mariner 9 succeeded in obtaining a global map at about 1 km resolution—similar to the resolution on the Moon obtained with ground-based telescopes.

Mariner 9 provided our first global view of another planet, and the results were spectacular. Revealed was a varied landscape of great volcanoes, deep canyons, and desert sand dunes, as well as the ubiquitous impact craters seen by Mariner 4. Most exciting were indications of ancient river beds carved by running water. Clearly, a 1-percent sample of a planet can be deceiving. (Imagine, for instance, the differing concepts of the Earth traditionally held by the nomads of the Sahara, the hunting cultures of the Arctic, or the island peoples of the Pacific.) We will describe this global perspective in a moment. But first let us look at the most successful of all Mars missions.

Viking

Launched to arrive at Mars on the two-hundredth birthday of the United States, Viking was the most ambitious planetary mission attempted to that time, consisting of two orbiters and two landers. Its primary objective was to answer a fundamental question that everyone could understand: "Is there life on Mars?" Earlier missions had revealed that the climate of Mars is extremely harsh by terrestrial standards, but no one was willing to rule out the possibility of microbial life. If terrestrial life forms could survive and grow in environments that range from boiling hot springs to the dry valleys of the Antarctic, who can deny the possibility of a native martian biota? Even though Viking failed to find life on Mars, this question persists.

The first Viking spacecraft orbited Mars on June 19, 1976 and began the search for a safe landing site. Touchdown was planned for July 4, but Mars did not cooperate. The target identified on the basis of Mariner 9 photographs turned out, when viewed by the superior camera of the Viking orbiter, to be unsuitable. No one wanted to risk a landing failure, the likely result if the spacecraft came down in a boulder field, but individual boulders could not be photographed from orbit. Therefore the Viking geologists had to use indirect evidence to infer the presence or absence of rocks on the surface.

In an atmosphere of mounting tension and excitement, daily meetings of the Viking scientists and managers were held to review the data and assess their interpretation. Finally a site was chosen, and on July 20 the Viking 1 lander fired its retrorockets, just seven years after Neil Armstrong's historic first step on the Moon. The lander spacecraft made a fiery entry through the thin martian atmosphere and softly touched down on the desolate plains of Chryse Planitia –

The surface at both Viking landing sites consisted of numerous rocks lying in a compact wind-deposited soil of red clay. Similar dust suspended in the atmosphere gives the sky a pale pink hue. Recently reprocessed images, shown here and on page 122, represent our best representations of the appearance of the martian surface.

This wide-angle view of the martian surface was transmitted by a Viking lander.

the Plains of Gold. A month later the second Viking landed successfully in a place called Utopia.

The Viking mission continued for six years, returning vast quantities of data to Earth. While the two landers directly sampled their environment and carried out their search for martian microorganisms, the orbiters surveyed the planet from above. Late in the mission the orbiters were brought closer to the surface for improved images, achieving resolutions as great as 10 m in some parts of the planet, 10 times better than the Magellan images of Venus. These images could distinguish features the size of a house or a small river, although unfortunately neither houses nor rivers were found on Mars. The Chryse lander was still functioning on November 5, 1982, when an operator error misdirected its antenna and permanently broke the communications link with the Earth. At the time it sent its final signals, the lander was serving as a weather station, providing information for daily martian weather forecasts.

Most of what we know about Mars is a result of the Viking mission. The Soviet Union sent spacecraft to Mars during the 1970s, including both orbiters and landers, but they were prone to failures and added little to our knowledge of the red planet. As for the United States, it turned its back on Mars after Viking. Perhaps

Viking was too successful; certainly it set a high standard for succeeding spacecraft, and it was clear that a great deal of money would be needed to achieve a major improvement. Automated rover missions were studied, as were missions to return samples of martian soil and rock to the Earth, but these came to nothing. Not until 1992 was the next Mars mission launched—a rather simple orbiter called Mars Observer.

Perhaps the problem was more fundamental. Viking was sent to Mars to look for life, and it found none. The idea that Mars truly is a dead world takes some of the glow out of its continuing exploration. We will return to the search for life on Mars at the end of this chapter, after we describe the planet as revealed by Viking data.

Global Perspective

Mars is a small planet. Its mass is just 11 percent that of the Earth, and its diameter of 6787 km is little more than half the diameter of our own planet or Venus. Being less massive, it also has a lower surface gravity, only about one-third that of Venus or Earth. Still, Mars is substantially larger than the Moon or Mercury.

Based on its size and surface gravity, we might expect Mars to be intermediate in many properties between Earth and Moon. Its atmosphere should be less than the Earth's but greater than the Moon's; similarly, its level of geologic activity should lie between the active volcanism and plate tectonics of the Earth and the dead, cratered surface of the Moon. This is just what we see on Mars.

As we have noted in looking at the other terrestrial planets, a measure of geologic activity is provided by the numbers of impact craters. A planet with little or no internal activity will be heavily cratered like the Moon. One very active, like the Earth, will display hardly any craters. Roughly half the surface of Mars is moderately heavily cratered, intermediate in number between the lunar highlands and the maria. These craters are eroded by windblown dust, but they can still be counted and their numbers compared with those on the Moon. This cratered terrain constitutes the oldest part of the surface,

The Argyre impact basin, the second largest impact feature on the planet at 700 km in diameter, was photographed by the Viking orbiter. The mountains that ring the basin were formed by the impact. Note also the layers of haze in the martian atmosphere.

and it must date back several billion years, at least as old as lunar maria.

In contrast, other parts of Mars have few craters. These areas must be relatively young, although most are not as young as the widespread plains of Venus. The terrain with fewer craters is volcanic in nature, as on Venus; apparently volcanic eruptions have flooded many ancient parts of the surface. While the level of volcanism is lower than on Venus or the Earth, it is still quite substantial, and it indicates a persistent activity that may continue today.

The distinction between the older, more cratered areas of Mars and the younger volcanic terrain is reflected in the surface elevations.

Since there is no liquid water, we cannot speak of a sea level elevation, but astronomers have arbitrarily identified a level that corresponds to the lower-lying plains, and they measure other elevations with respect to this designated level. It turns out that all of the older, cratered areas are at high elevations, while many of the volcanic plains are depressed. There is a global pattern to these differences, with the ancient uplands occupying one hemisphere (roughly the southern hemisphere of the planet) and the younger lowlands the opposite hemisphere. A clearly identifiable boundary separates these areas, characterized by a change in elevation of several kilometers.

This global division of Mars into uplands and lowlands is one of the fundamental mysteries of the planet. If the uplands represent the original crust, what happened to this crust in the younger areas? If it had simply been flooded by lava, it would still lie at high elevation. Something apparently stripped away about 5 km of the crust previous to the lava inundation. Since there is no place for the excavated material to have gone, there must be a flaw in this analysis. More likely, the crust in the northern hemisphere collapsed in response to internal forces acting billions of years ago, but the details of this process are not understood.

In addition to the hemispheric dichotomy represented by the ancient uplands and the younger lowland plains, there is one large continent that straddles these two terrains and overlies them both. This is the Tharsis bulge, a huge uplifted continent about the size of North America, crowned by the highest volcanoes on Mars. Tharsis combines uplift driven by tectonic forces with volcanism that has piled still more lava on top of the uplift. Here we have the unusual situation of lava pouring out at high elevations instead of filling in lowlands — the common situation on Earth, Venus, Mercury, and the Moon.

The Tharsis bulge with its volcanic topping represents the most recent major adjustment of the mantle and crust of Mars. By carefully studying its structure and counting the impact craters superposed on its surface, geologists have concluded that the Tharsis continent formed subsequent to the hemispheric division between uplands and lowlands. Many interpret Tharsis as the result of a large mantle plume, possibly analogous to the activity that some scientists think maintains the Maxwell Mountains on Venus.

The large-scale topography of Mars is quite different from that of either Earth or Venus, as can be seen when this image is compared with those on page 90. Note the hemispheric dichotomy between the southern uplands and the northern lowlands. The white bulge is Tharsis, and the prominent circular depression in the uplands is the Hellas impact basin.

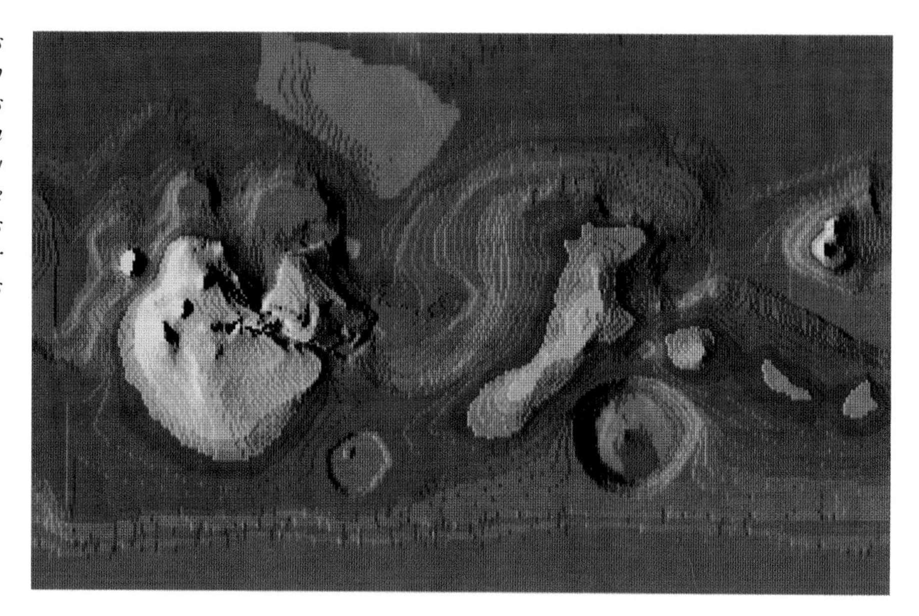

Mars, like Venus, is a planet that has been subjected to many tectonic stresses; moreover, in both cases these forces are insufficient to drive the crust sideways, initiating plate tectonics. Mars is another planet with so-called blob tectonics, but there is just one blob, apparently less active than its counterparts on Venus.

Recall that our understanding of the Moon was greatly advanced by laboratory analysis of returned lunar samples. It has only recently been recognized that we also possess a few samples from Mars, in the form of the SNC meteorites. (The initials abbreviate a series of technical terms in meteorite nomenclature, based on the locales of discovery sites.) These martian samples have come to us *gratis* as ejecta from one or more large craters on Mars. The SNC meteorites, about a dozen of which have been found (mostly in the Antarctic), are composed of basalt and apparently originated from cratering of relatively recent volcanic flows, perhaps in the Tharsis region. Laboratory analysis of these meteorites is providing important information on the composition of the martian crust and mantle, but unfortunately we do not know the location or geologic context in which these samples formed, so interpretation is difficult.

Tharsis with its crown of recent volcanoes proves that Mars is not geologically dead. However, we actually know very little about the interior of the planet. The Viking landers did not determine the rate of heat flow from the interior, nor did they detect any marsquakes. We infer from the SNC meteorites that the interior structure and composition are similar to that of the Earth, although the core probably contains more sulfur than our own (which is mostly iron) and is likely to be solid, since Mars has no measurable magnetic field. Apparently it is a planet slowly running down as its interior heat escapes.

On the Surface

While humans have never set foot on the surface of Mars, we have sent our robotic stand-ins to take a look around for us. In previous generations, it was assumed that space exploration would proceed as had that of the Earth, with brave men leading the way to new worlds. However, the revolution in electronics, communications, and

The Tharsis region of Mars includes both volcanic and tectonic features. The two large shield volcanoes shown in this Viking photograph are about 50 km across. The numerous tectonic cracks are the result of tension in the crust, probably related to the uplift of the Tharsis region.

computers has opened up remarkable opportunities for an exploration partnership between humans and machines.

The Viking and Voyager spacecraft were among the first machines to take advantage of these new technologies. We have already seen how the reprogrammable computers and autonomous operating systems of Voyager enabled the grand tour of the outer solar system. The two Viking landers, designed and built at the same time as Voyager, played a similar role in the exploration of Mars.

The Viking 1 landing site at Chryse Planitia is at 22 degrees north latitude, in a windswept plain near the lowest point of a broad basin. From orbit, the Chryse Planitia can be characterized as a 3-billion-year-old lava plain in the martian lowlands, with indications of possible subsequent flooding by water. This information did not fully prepare us, however, for the actual view on the surface.

Shortly after touchdown, the Viking 1 lander activated its twin cameras. The pictures transmitted to Earth showed a desolate but strangely beautiful landscape of reddish-brown rocks and soil under a cloudless pink sky. Numerous angular rocks up to a meter across lay partly buried by fine-grained soil, much of it sculpted into dunelike shapes, and on the horizon low hills apparently marked the rims of distant impact craters.

Viking 2 landed further north, at a latitude of 48 degrees. The view was generally similar at this site in Utopia, but there were substantially more rocks: Viking 2 apparently landed on ejecta from a large crater called Mie, about 200 km away. Both landscapes appear to be in a state of erosion rather than deposition, with the winds gradually stripping away the soil to reveal rocks beneath.

In addition to photographing its surroundings, each lander sniffed at the atmosphere with a variety of analytical instruments and poked at nearby rocks and soil with its mechanical arm. As part of its primary mission to search for life on Mars, each Viking collected soil and analyzed it for microbes, as we will describe later. The soil was found to consist of clays and iron oxide, as had been expected from its color.

Each lander deployed a weather station on a boom 1.3 m high to measure temperature, pressure, and wind speed and direction. The average surface pressure was less than 1 percent that of Earth, as had been inferred from remote measurements, but this pressure varied substantially over time and season. Temperature excursions were large, in the absence of moderating atmosphere or oceans. Typically, the summer maximum at the tropical latitude of Viking 1 was −33 C, dropping to −83 C just before dawn. The lowest air temperature, measured further north by Viking 2, was −100 C. During the winter, Viking 2 also photographed frost deposits on the ground.

Most of the winds measured at the Viking sites were low, typically from about 2 m/s at night up to 7 m/s in the daytime. Mars is capable of great windstorms, however, which can shroud the planet

At the northern Viking landing site in Utopia, atmospheric water can freeze on the surface to produce white patches of frost. This early morning view was obtained in late winter.

in dust; during such storms the sky turns dark red and the Sun is greatly dimmed.

The Viking landing sites were not typical of the planet; indeed, they had been selected specifically because they were dull and safe. Proper interpretation of the lander data requires a broader perspective on the surface of Mars, obtainable only from orbital surveys.

Volcanoes and Canyons

The Mariner 9 orbiter, which made the first global survey of Mars, began its photographic mission during a global dust storm that obscured the planet's surface. When the dust began to settle a few weeks later, the first features that appeared were four dusky spots, each with a large crater in the center. What could this mean? We knew from previous flyby missions that impact craters were present, but it seemed an extraordinary coincidence that these spots should each have a central crater.

The solution to this mystery transformed our conception of Mars as a dead, Moon-like world. Viking scientist Harold Masursky of the U.S. Geological Survey reasoned that the four dark spots must be the four highest mountains on Mars, since they were the first features to appear above the swirling dust. And the four highest mountains were topped with craters. Surely, then, the craters must be associated with the mountains – instead of impact craters, these are volcanic craters (calderas). Further, the four volcanoes must be huge features, larger than the tallest terrestrial volcanoes.

As the dust continued to subside, Masursky's hypothesis was confirmed. As each volcano emerged, we could trace the ancient lava flows that extended down its flanks. Three of these volcanoes lie along the high ridge of Tharsis. Nearby, the fourth of the great volcanoes, Olympus Mons, is the largest such feature in the solar system. Olympus Mons is a shield volcano like those of Hawaii but larger, 16 km in height and nearly 500 km across. Its summit caldera is large enough to hold the entire Hawaiian island of Oahu.

Is Olympus Mons still active? It is impossible to tell. There are no impact craters on its flanks or in its summit caldera, indicating that the surface we see must be younger than a few tens of millions

Olympus Mons is the largest volcano on Mars and probably the largest in the solar system. In this view from orbit, the 25-km-high shield volcano is surrounded by clouds. The multi-floored central crater, or caldera, is about 80 km in diameter.

of years. Therefore, we can assert that Olympus Mons is young on a geologic time scale. Perhaps intermittent eruptions continue to occur, but there is no way to determine this unless we should happen to spot the "smoking gun" of an actual eruption.

We can also ask why Olympus Mons is so much larger than the volcanoes of Venus or Earth. Two factors contribute to its size. First is the matter of lifetime. It appears that the Tharsis bulge formed more than 2 billion years ago, and its large volcanoes may have been active over a comparable span of time. In contrast, the oldest volcanoes on Earth have ages of just a few million years. The geology of our planet is constantly transformed by plate tectonics; even when a long-lived hot spot develops, as it has under the Hawaiian islands, the crust keeps moving with respect to the source of heat. In the case of Hawaii, the result is a long chain of islands, formed as the crust creeps over the mantle at a speed of several meters per century. Conditions are more static on Mars, and volcanoes can persist there much longer.

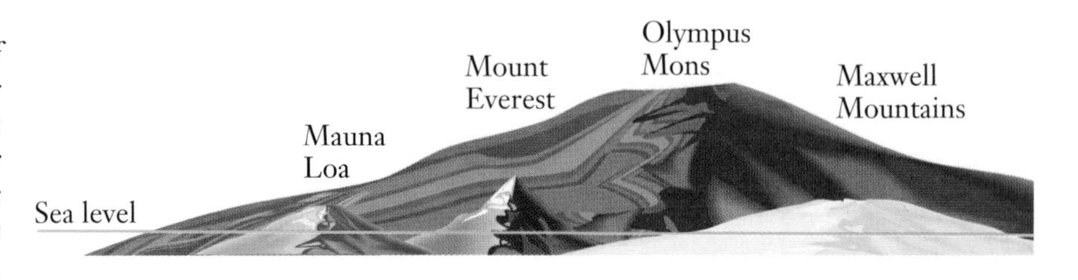

The largest mountains on Mars dwarf their counterparts on the Earth or Venus. Olympus Mons on Mars has a volume more than 20 times greater than Mauna Loa in Hawaii. Mountains can grow larger on Mars in part because of its lower surface gravity.

The difference in gravity between Mars and the Earth or Venus also contributes to the size of Olympus Mons and the other martian volcanoes. The maximum height of a volcano on any planet is set by the ability of the rock to support the weight of the mountain. Mauna Loa, which is currently active, illustrates this point. As new lava erupts near the summit, the added weight causes the entire mountain to slump outward; consequently, its altitude does not change. The same effect holds on Mars, but because the force of gravity is only about one-third as great, martian volcanoes can grow more than twice as tall as their terrestrial counterparts before they reach their limiting heights.

In addition to the four large Tharsis shield volcanoes, Mars has hundreds of smaller shields, concentrated primarily in the general area of Tharsis and in the older volcanic region of Elysium. There are smaller cinder cones as well, down to the limiting resolution of the Viking photographs. Most of the younger lowland basins seem to be huge lava plains, similar to the lunar maria. In its volcanic areas, the level of volcanism on Mars approaches that on Venus. Unlike Venus, however, Mars has vast tracts of older uplands where no volcanic eruptions have taken place.

The same internal forces that uplifted Tharsis generated tremendous tectonic stresses in the crust of Mars, causing it to stretch and crack. One result is a system of canyons that stretch for nearly 5000 km from the slopes of Tharsis out across the old uplands. This system — called the Valles Marineris, after Mariner 9, the spacecraft that discovered it — is about 4 km deep and up to 100 km wide. Even its smaller tributaries dwarf their terrestrial counterparts.

We call the Valles Marineris a canyon, but the term is misleading. On Earth, canyons are steep-walled valleys cut by

The Valles Marineris is a complex system of tectonic canyons. The valleys have been widened by landslides that leave characteristic scalloped walls with aprons of debris on the canyon floor.

running water. The Valles Marineris is indeed a steep-walled valley, but water did not carve it; instead it is akin to a giant crack, the result of tectonic forces associated with the Tharsis uplift. You can tell this by the remarkable straightness of the Valles Marineris, in contrast with the sinuous course of the Grand Canyon of the Colorado in Arizona or the Grand Canyon of the Verdon in Haute-Provence. No river ever flowed down the Valles Marineris, nor is there any outlet from these canyons to the lowland basins to the north.

Although the Valles Marineris originated as a series of large, parallel cracks on the slopes of the Tharsis bulge, these cracks have subsequently widened and coalesced to achieve their present width. Here water played a role. From their structure, geologists have concluded that the steep walls were formed through undercutting by subterranean springs and subsequent collapse; landslides associated with recent collapses can still be seen in the valley floor. Fine dust was then scoured away by high winds and deposited elsewhere, leaving the wide, deep canyons we observe today. This entire process required more than a billion years, 100 times longer than the time the Colorado River has been grinding away at its Grand Canyon on Earth.

Atmosphere

The great dust storms are the most obvious evidence for a martian atmosphere. Smaller white clouds can also be seen, sometimes in association with the large volcanoes, which attract clouds just like the mountains of our own planet. Olympus Mons acquired its name from a white spot visible from the Earth more than a century ago and called Nix Olympica, or Snows of Olympus. Now we know the white spot was not snow but clouds that form over the volcano from time to time.

The actual quantity and composition of atmosphere on Mars was a matter of speculation until rather recently. Early in the present century, astronomers guessed that the martian atmosphere was mostly nitrogen, like that of the Earth. At the dawn of the space age, a sort of consensus had been achieved that the surface pressure was about 8 percent that of the Earth, or 0.08 bars, but this was based on highly indirect reasoning. Mariner 4 provided the answer in 1964, when it passed behind the planet and its radio signal penetrated the martian atmosphere in order to reach the Earth. From the manner in which the signal was altered, we determined that the average lowland surface pressure was much lower than predicted: just 0.007 bars, or less than 1 percent that of the Earth.

By combining the spacecraft measurement of the surface pressure on Mars with ground-based spectra of the planet, astronomers quickly concluded that carbon dioxide is the primary gas. Nitrogen is a minor constituent, while oxygen is virtually absent. Oxygen detection would have argued for the possibility of photosynthetic life on Mars, since oxygen is unstable in a planetary atmosphere without a continuing source.

In the absence of oxygen there is no ozone, so that solar ultraviolet light penetrates to the martian surface. Unattenuated ultraviolet light is lethal to exposed microorganisms and results in an environment difficult at best for the survival of life on Mars.

The Viking landers made detailed direct measurements of the composition of the martian atmosphere. Results are shown in the table, which compares the atmosphere of Mars with that of Venus and the Earth. In spite of the tremendous difference in total quantity and a surface pressure on Venus more than 10,000 times greater than

Composition of the Atmosphere of Venus, Mars, and Earth (Percentage)

Gas	Venus	Mars	Earth
carbon dioxide (CO_2)	96	95	0.03
nitrogen (N_2)	3.5	2.7	78
argon (Ar)	0.006	1.6	0.9
oxygen (O_2)	0.003	0.15	21

on Mars, each planet's atmosphere is roughly 96 percent carbon dioxide and 3 percent nitrogen. Today we tend to associate this composition with the primitive atmosphere of the Earth as well.

Three kinds of clouds can form in the martian atmosphere. We have already mentioned the occasional dust storms, analogous to those generated in the Sahara or Gobi deserts of the Earth. Since Mars is all desert, there is more opportunity for dust to be raised by high winds; on occasion the dust storms can reach global dimensions. Second, there are low clouds or fog composed of water, again similar to those on the Earth. Water clouds are much less common on Mars, however, because the planet is very dry, and they tend to form only at night or in the winter. The third type of cloud, unique to Mars, is composed of condensed carbon dioxide or dry ice. Such clouds can form only when the atmospheric temperature drops below the freezing point of CO_2 at about -125 C. On Earth it never gets that cold. On the whole, the atmosphere of Mars is much clearer than that of our planet.

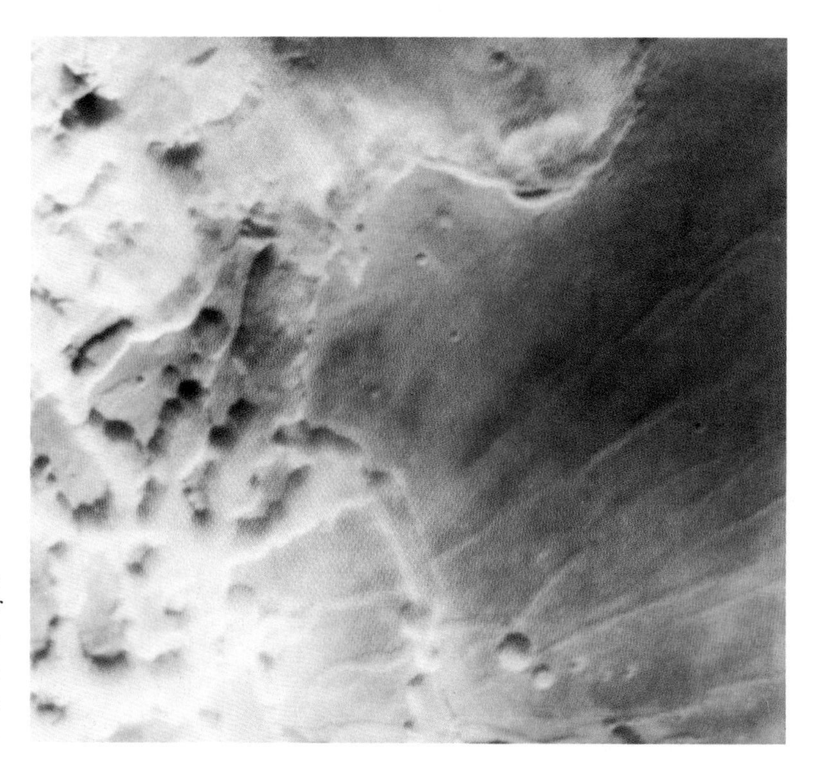

Even on the dry Mars of today, low fog can form by the condensation of atmospheric water vapor. This orbiter view shows morning fog in the canyonlands. The width of the frame is about 1000 km.

Mars has its weather, which can be forecast for that planet just as it is for the Earth. The primary force that drives weather on any planet is the differential heating of the surface by sunlight. The atmosphere responds to try to equalize temperatures, transporting heat from day to night and from summer to winter. On Mars the daily temperature variations are very large and the winds proportionately fierce.

Much of the motive force for the weather comes from the transport of both carbon dioxide and heat between the north and south polar regions. During the winter, atmospheric CO_2 can condense at the cold pole and freeze onto the surface, significantly lowering the total surface pressure of the atmosphere. This CO_2, released with the coming of spring, flows from the pole toward the equator and ultimately to the opposite hemisphere, when it again condenses. This seasonal pressure variation was directly measured by the Viking landers. Nothing quite like it happens on our own planet.

Polar Caps

The polar caps of Mars play a role in both the daily weather patterns and the long-term global climate. There are three kinds of polar caps, and we must be careful to distinguish them.

When we look at Mars through a telescope, the polar caps we usually see are the seasonal surface coating of carbon dioxide frost that forms whenever the temperature drops below about −125 C. This dusting of dry ice is analogous to the snowfalls that blanket both temperate and polar regions of the Earth during the winter. Like the terrestrial snow cover, the martian polar caps extend down to about 45 degrees latitude at midwinter and then retreat toward the pole in spring.

The composition of the dry-ice polar caps has never been measured directly, but we infer it from their temperatures. As the major component of the atmosphere, carbon dioxide will freeze whenever the temperature drops low enough. Similarly, the retreating edge of a melting dry-ice cap will be at precisely the freezing temperature of −125 C, just as a melting ice cube is always at exactly

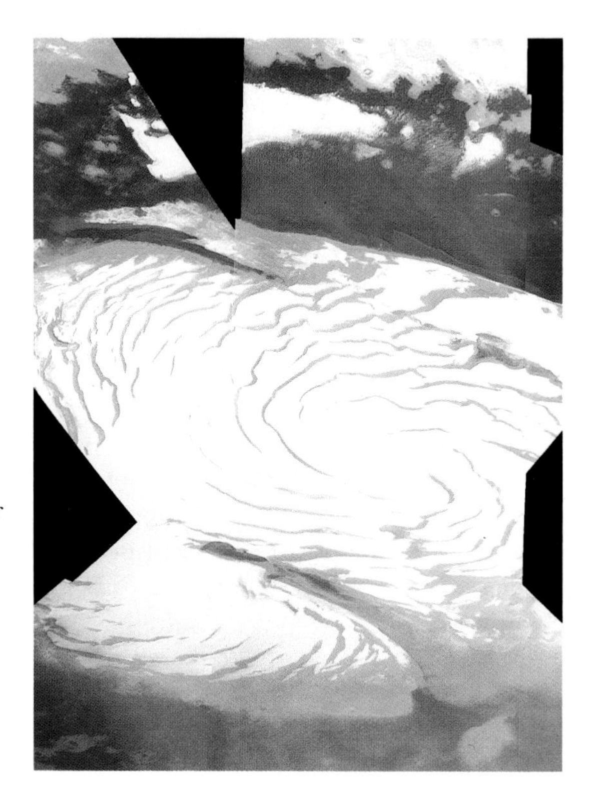

The residual north polar cap of Mars is about 1000 km across. In midsummer the CO_2 frost disappears to reveal this underlying cap of H_2O — the primary surface outcrop of a large subsurface reservoir of water ice believed to exist on Mars.

the freezing point of water. Thus temperature measurements of the caps at various seasons, which can be made from orbit, confirm their composition.

At the south pole of Mars the seasonal dry-ice cap never entirely disappears. A remnant polar cap with a diameter of about 350 km persists through the summer's warmth. This permanent deposit of dry ice, presumably mixed with frozen water, is the second kind of martian polar cap.

The third type of polar cap, composed of ordinary water ice, is perhaps the most interesting. In the late northern summer, the dry ice evaporates completely to leave a residual northern polar cap nearly 1000 km across, composed of water ice. We know there is no dry ice here because the measured temperatures are too high for CO_2 to remain solid; in addition, the water vapor in the atmosphere is enhanced over this pole, as would be expected if it were composed of sublimating water ice. There are no really adequate theories to explain the difference between the two poles.

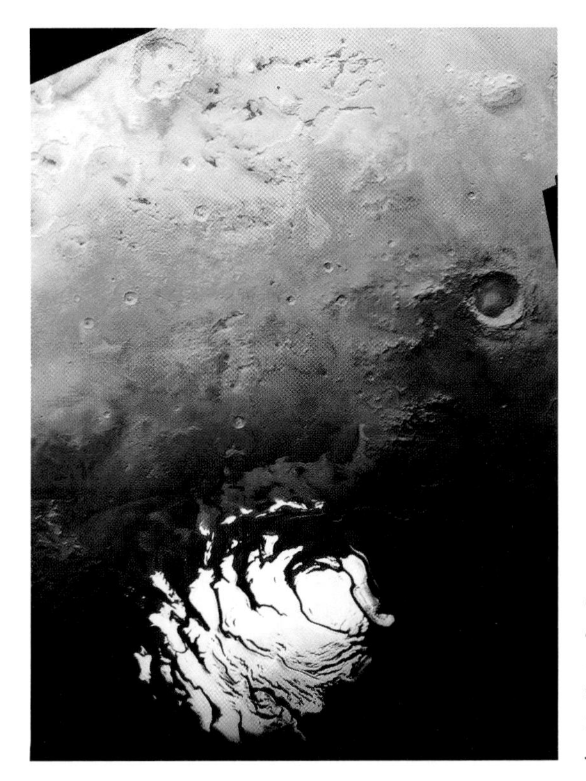

The residual south polar cap is composed of a mixture of CO_2 *and* H_2O *ice. The cap is about 350 km in diameter and remains at a very low temperature through the summer months.*

The remnant water-ice cap at the north pole and the presumptive water ice that persists at the south in association with the remnant dry-ice cap are the most visible portions of a much larger quantity of frozen water that is thought to exist on Mars. The two caps are the tip of a subsurface H_2O iceberg, with the rest in the form of permafrost. Unlike Venus, Mars probably retains much of its original ocean of water, but that ocean is solidly frozen beneath the surface. Calculations suggest that the equatorial regions are free of ice up to a latitude of about 40 degrees. From that latitude to the poles, a substantial part of the crust could consist of frozen H_2O. Only near the poles does this layer of ice emerge.

The southern polar cap of Mars has another fascinating story to tell us, although it probably has nothing to do with the presence or absence of oceans of ice. When photographed in detail from orbit, the area around the pole is seen to consist of plates, or laminae, a few meters thick. When exposed on eroded slopes, these plates form easily visible terraces. Apparently there are alternating layers of dark and light sediment throughout the polar regions. The high-latitude parts of Mars are generally areas of deposition, with extensive sand dunes as well as the layered polar deposits. Probably these represent the final resting place of most of the dust stripped from the Valles Marineris and other eroded lowlands.

The polar layered deposits are so extensive and so regular that they must be the product of some kind of cyclic or periodic climate change. It is estimated that no more than a few millimeters of dust could be deposited each year at the poles, so layers tens of meters thick must correspond to depositional time scales of tens of thousands to hundreds of thousands of years. Thus the martian polar regions provide evidence for periodic changes that recur at intervals similar to those between the great ice ages on Earth. Understanding of the periodic climate changes on Mars may hold the key to the interpretation of the past climate of our own planet as well.

The layered deposits contain clues to the relatively recent past of Mars. However, they are no more than a few million years old, providing no information concerning the long-term evolution of the atmosphere and surface. For such a perspective, we must return to the geology of the planet as revealed from orbiting spacecraft.

Ancient Rivers

The most amazing accomplishment of Mariner 9 was the discovery that the ancient climate of Mars was more clement than that of today. The evidence comes from the presence of many ancient river beds that could only have been formed by the erosive action of liquid water. These dry river channels, found primarily in the old uplands of the planet, date back 3 to 4 billion years.

The martian channels discovered by Mariner 9 should not be confused with either the canyons, like Valles Marineris, or the fictitious canals of Percival Lowell. Neither tectonic in origin nor long, straight lines like the canals, they are also unlike the lava channels seen on Venus and the Moon. Judging by their appearance as photographed by the Mariner and Viking orbiters, they really were carved by running water.

Runoff channels in the ancient uplands provide evidence that rain-fed rivers once flowed on Mars. The width of this Viking frame is about 200 km.

There are two types of martian channels, and each has a different story to tell. We begin with the older runoff channels of the martian uplands, which most resemble terrestrial river systems. Many runoff channels are simple valleys, tens to hundreds of meters wide and a few tens of kilometers long. Others, more interesting, are extensive networks of interconnected channels like the dendritic stream networks of the Earth: small tributaries connect into larger channels to provide drainage for a substantial area.

Some of the runoff channels may have originated in underground springs, but others clearly indicate the presence of multiple tributaries fed by rainfall. They tell us that Mars once supported free-flowing rivers and experienced the miracle of rain. When did this take place? From the fact that the runoff channels are restricted to the older uplands, we can date their existence to the period before the formation of either the lowlands or the Tharsis continent. This

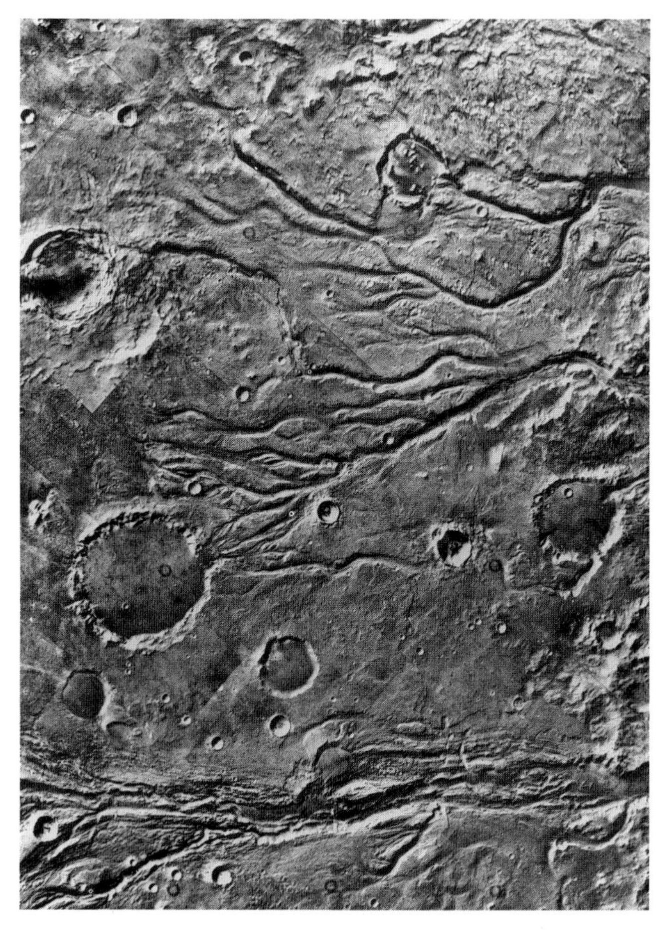

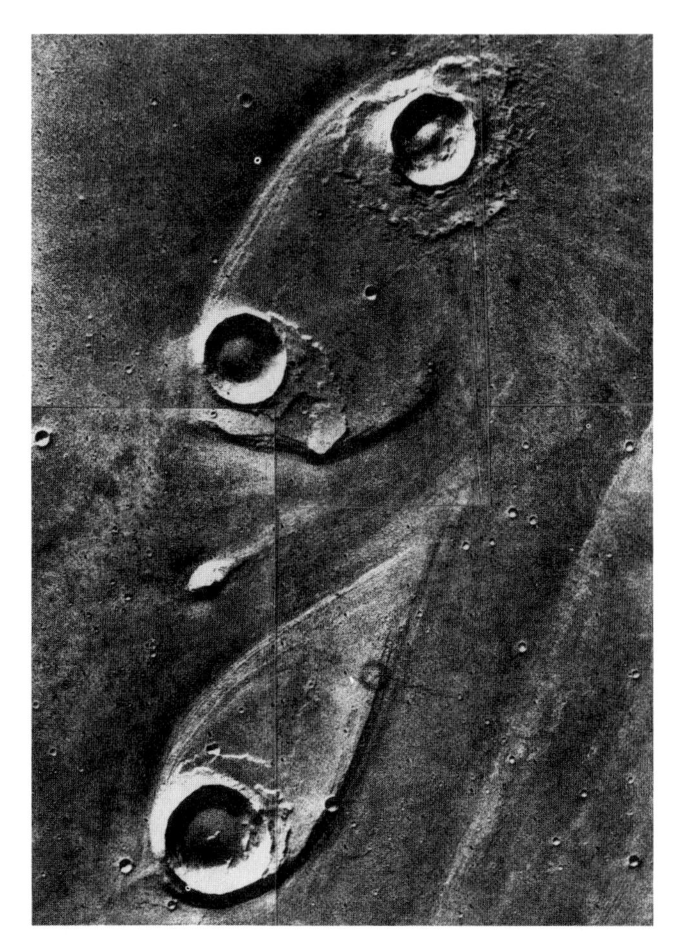

Large outflow channels that once carried massive floods of water are a unique feature of martian geology. The channels shown in the frame on the left, which has a width of about 400 km, drain from the uplands into the Chryse basin, where the Viking 1 spacecraft landed. The image on the right is a close-up of teardrop-shaped islands near the terminus of one of the major channels.

places their age back about 4 billion years, to the period of the late heavy bombardment on the Moon. This is also about the time that life first arose on the Earth.

Larger and more controversial are the second class of martian channels, called the outflow channels. These are individual broad rivers without tributaries, flowing from the equatorial uplands down into the lowland plains. Like the smaller runoff channels, the outflow channels appear to have been carved by running water. Flows in the outflow channels were probably intermittent and catastrophic, however, and the source of the water was not rainfall.

The largest outflow channels drain into the Chryse basin where Viking 1 landed. Half a dozen of these channels are 10 km or more in width. They originate on the Tharsis slopes and drop several kilometers in altitude as they descend into the basin. Over a length of several hundred kilometers they have cut multiple parallel channels that diverge and interconnect, displaying characteristic patterns of water erosion such as streamlined islands and sand bars.

The scale of the martian outflow channels dwarfs terrestrial rivers. It is estimated that any one of these channels could have supported a flow rate 100 times that of the Amazon River on Earth. It is difficult to imagine sustained water flow of this magnitude on Mars. Rather, geologists attribute the formation of outflow channels to events similar to terrestrial floods caused by the sudden failure of natural dams, such as those that carved the Grand Coulee and other dry channels in eastern Washington during the last ice age.

The most likely source of such catastrophic floods on Mars was the melting of subsurface permafrost at the time of formation of the Tharsis continent about 3 billion years ago. If ice deposits existed in tropical latitudes at that time, volcanic heating might have released the water in dramatic, episodic floods. Perhaps each such flood lasted only a few days or weeks. Maybe the water quickly evaporated or refroze at higher latitudes. Or possibly temporary ice-covered lakes or seas formed in basins like Chryse. We do not know.

Climate Change

Several lines of evidence point to a martian past very different from the cold, dry planet we see today. The terraces of the martian poles tell us of cyclic climate change over the past few million years, although probably these changes represent rather small variations. But if we turn the clock back 3 billion years, we see indications of huge, episodic releases of water from melting permafrost, perhaps generating ice-covered seas in the lowland basins. At 4 billion years, there is persuasive evidence for warmer temperatures, falling rain, and flowing rivers.

It is interesting that the evidence for climate change on Mars is more complete and compelling than it is for our own planet; if we are interpreting this evidence correctly, Mars has suffered greater changes than has the Earth. In addition, however, we have the advantage that we can more easily probe into the past on Mars, whose surface reveals terrain that dates back 3 and 4 billion years. On the Earth, our more active geology tends to erase evidence of conditions more than a few hundred million years old. Even though we have only a cursory overview of Mars obtained from orbiting spacecraft, we can outline with some confidence a broad span of climate history on that planet.

If Mars was once warmer than it is today, the most likely explanation lies in an enhanced atmospheric greenhouse effect. During the first half-billion years of its history, Mars presumably had a denser atmosphere of carbon dioxide, carbon monoxide, and water vapor — similar to that of the primitive Earth and Venus. If the pressure of that atmosphere were at least several bars, the resulting greenhouse effect could have sustained surface temperatures above the freezing point of water. While we cannot prove the existence of these conditions on Mars, there is no reason to exclude the possibility of such an atmosphere and widespread liquid water during the first few hundred million years of martian history.

Because Mars is farther from the Sun than the Earth, it has always tended to be cooler. Further, the lower surface gravity on Mars makes it harder for that planet to maintain its initial dense atmosphere. Some combination of less solar energy and the escape of the atmosphere thus led to a progressive degradation of conditions. Presumably the 4-billion-year-old runoff channels of the equatorial uplands record a stage in that evolution when the primitive seas may have evaporated or frozen but rain still fell in tropical latitudes.

By the time internal geologic activity generated the great floods that formed the outflow channels, the atmosphere of Mars might already have shrunk to nearly its present state. Perhaps these catastrophic floods injected more water vapor and briefly raised the surface temperature again, but they could not reverse the long-term cooling of the planet. Mars would then have experienced a sort of runaway refrigerator effect, the opposite of the runaway greenhouse

Current evidence suggests that Mars has experienced dramatic changes in climate, as indicated in these paintings. The first image (above left) shows the way the planet might have looked about 3 billion years ago, if plentiful liquid water was available at the time of formation of the Tharsis bulge and the outflow channels. In the next image (above right) most of the liquid water is gone but substantial deposits of ice, and perhaps of ice-covered lakes, remain. The final image (left) shows the planet as it is today.

effect sometimes suggested as the origin of the high temperatures on Venus. Escape of the atmosphere led to surface cooling, which in turn resulted in freezing of water and a further drop in temperature. The result is the frigid, desiccated world we see today.

The Search For Life

Mars may now be uninviting, but its past history suggests that conditions were once suitable for the origin and proliferation of life similar to that on the Earth. If life once existed, was it able to adapt to the progressive martian climate change? Or are conditions today simply too harsh for any life to survive? We do not know. In any case, Mars is the best that nature has to offer as an alternative abode for life in the solar system.

The idea of life on Mars is not new. As soon as astronomers of the Renaissance realized that the planets were worlds like the Earth, they began to speculate about the possibility that these worlds were inhabited. It is surprising to realize, from our modern perspective, that 300 years ago it was widely assumed in Europe that all planets were inhabited by intelligent creatures. Even as evidence accumulated to contradict this optimistic hypothesis, Mars seemed the exception.

Early in the twentieth century the work of Percival Lowell and his contemporaries seemed to indicate the presence of life on Mars. The canals provided the strongest evidence, and when they faded the hopes of intelligent life faded with them. However, indications of plant life or at least of microorganisms—indirectly deduced from the colors and observed seasonal changes on Mars—continued to stimulate both astronomers and the public up to the dawn of the space age. Unfortunately, the interpretation of this evidence was driven more by wishful thinking than hard analysis.

The discovery of an independent, indigenous life form on another planet would be one of the most important products of modern science. The current explosion in our understanding of the molecular basis of life and heredity still begs the fundamental question of the uniqueness of life. All life on Earth is related to all other life, and presumably we all arose from a single ancestor. From *E. coli* to elephants, we are all made of the same genetic material, packaged in a variety of ways. Are other life forms possible? If so, are they based on DNA or RNA or similar molecules? What in our particular biochemistry is essential and what is simply the incidental product of the way life originated on Earth? If we found life on Mars, we could begin to answer such questions. Thus it is no surprise that the first martian landers took the search for life as their most important objective.

How should we search for life on Mars and are we confident that we could recognize martian life if it were present at the Viking landing sites? The chemistry of the martian atmosphere argues against widespread life similar to that which has so altered our own atmosphere, and the low temperatures and ultraviolet radiation at the surface suggest that Mars is not teeming with life like our own planet. No one expected the Viking cameras to reveal waving palm trees or exotic insects scurrying past the spacecraft. Viking concentrated instead on a search for martian microorganisms, which may have found microenvironments in which they could escape some of the rigors of the martian climate.

Viking carried out five experiments that contributed to the search for life: two indirect and three direct. The indirect experiments consisted of detailed chemical analysis of the atmosphere and soil to search for organic molecules or other indications of the raw materials or by-products of life. The three direct experiments tested small soil samples for the presence of viable microorganisms.

Each lander contained a miniature biological laboratory to test the soil for evidence of three fundamental biological processes: respiration by living microorganisms, absorption of nutrients offered to any organisms that might be present, and exchange of gases between the soil and its surroundings. Soil placed in experimental chambers was incubated in contact with a variety of gases, radioactive isotopes, and nutrients (including liquid water) to see what happened.

The experiments that tested for absorption of nutrients and gas exchange showed some activity, but this was probably caused by inorganic reactions in the soil triggered by the presence of liquid water. However, the indirect experiments revealed no sign of organic material in the soil or of atmospheric gases that might be the product of biological activity. While the possibility of martian microorganisms

Evidence of the Viking search for life can be seen in this lander view, in the form of several trenches dug to collect soil for analysis. An ejected instrument cover also lies on the martian surface next to the spacecraft.

has not been eliminated by these results, most experts consider the likelihood of life on Mars to be very low. If life exists, it must be located somewhere quite different from the two sites sampled by Viking.

The negative results from Viking do not, however, prejudice the issue of an origin of life on Mars. Even if the planet is sterile today, there is no reason to doubt the possibility that life formed during the few hundred million years when the planet was relatively warm and wet. If so, there should be fossil evidence of life waiting to be discovered by future explorers in the rocks of Mars.

Does Mars Have A Future?

From a human perspective, Mars stands out as the one other potentially habitable body in the solar system. It is the only place where we may one day hope to live in relative safety and comfort, and the only place we are likely to find evidence for an origin of life independent of our own planet.

Whether or not Mars has ever been an oasis for life, it is a compelling target for in-depth exploration. Mars is a natural laboratory for studying the geologic forces that shape a planet, including volcanism, tectonics, and erosion by wind and water. The same is true for the study of planetary weather and climate. A better understanding of the weather of Mars could, for instance, teach us much about the weather of our own planet.

For all of these reasons, interest in continued investigation of Mars remains high. Before its disintegration, the U.S.S.R. had a long-term commitment to robotic and then human flights to Mars. On the twentieth anniversary of the Apollo landing on the Moon, the United States declared the goal of landing humans on Mars within 25 years. In spite of these statements of intent, however, it is not clear when humans will first take off on the long voyage to Mars. Many robotic precursor missions are required to learn more about Mars before we can commit to crewed missions, and the costs are bound to be high.

The Soviet Union is no more, and at this writing its space program is in disarray. Political controversy in the United States similarly blocks broad public support for the exploration of Mars. Some say that the age of exploration is past, that neither the United States nor the world can afford to reach out to another planet.

These doubts are legitimate, but despite them Mars beckons. If routine space travel is ever to become a reality, there must be a worthwhile destination—and that destination is Mars. As the author Ray Bradbury has eloquently asserted, there will someday be Martians, but these Martians will be explorers and colonists from planet Earth.

Future Mars missions will include robotic rovers to explore the surface. This is a Russian test vehicle shown operating in the volcanic terrain of the Kamchatka Peninsula. The rover, slightly smaller than a small automobile, is a "wheel-walking" machine, with independently driven conical wheels attached to a hinged frame.

FIRE AND ICE

Small Bodies in the Outer Solar System

•••

Jupiter's satellite Io is the most volcanically active object in the solar system; it is a prime target for study by the Galileo spacecraft, shown here at the time of its launch from the Shuttle.

There is much more to the solar system than the eight large planets and our Moon. Far from the Sun, low formation temperatures led to objects of very different composition from the terrestrial planets. We have already seen how the four giant planets grew so large that they were able to attract hydrogen and helium gas, with the result that they have no solid surfaces at all. In addition, however, numerous icy satellites and rings accompany the giant planets. The satellites are solid bodies that can be compared with the terrestrial planets, while the rings are made up of debris from shattered satellites. As we will see in this chapter, the smaller bodies in the outer solar system have many surprises for us.

Astronomers study satellites and rings primarily for the insight they provide on the formation of the solar system. The presence of these icy objects, perhaps remnants from the formation of the system, transforms each of the four giant planets into a miniature planetary system, which may mimic fundamental properties of the solar system itself.

Satellite Discoveries

Galileo Galilei, then a 46-year-old professor of mathematics at the University of Padua, constructed the first astronomical telescopes in 1609, experimenting with a variety of optical elements and testing the system to confirm that it accurately magnified distant objects. He was able to achieve a magnification of 30, but the telescope was difficult to point and use because of its tiny field of view. He began his astronomical studies in January 1610 by looking at the brightest objects in the sky: the Moon, Venus, and Jupiter.

On the first night of his observations, Galileo noticed that Jupiter was apparently accompanied by several small, faint objects. Over a period of a few hours these points of light stayed close to Jupiter but moved back and forth relative to the planet and each other. After a few nights of observation, Galileo was convinced that he had discovered companions that orbited Jupiter much as Copernicus had suggested the planets orbit the Sun. He had found the first planetary satellites, or moons—four small worlds that are still called the galilean satellites.

The galilean satellites provided a critical demonstration of the Copernican cosmology. Jupiter and its four moons form a solar system in miniature, with the satellites bound to the larger planet. In the traditional geocentric view this was impossible; gravity was thought to be a property of the Earth alone, and only the Earth could serve as the center of motion for astronomical objects. In contrast, Copernicus had suggested that the Sun was the center of attraction for the planets, just as the Earth was for the Moon. Now Galileo had discovered that Jupiter also was a center for orbital motion, a result inconsistent with the old geocentric theory. Copernicus had not predicted that other planets besides Earth might have moons, but the discovery of four jovian satellites, consistent with the Copernican cosmology, was taken as evidence in its favor.

While Galileo is credited with the discovery of the satellites of Jupiter, he was competing with another early telescopic observer, Simon Marius. Marius probably saw the four large jovian satellites independently the same year as Galileo, and he suggested individual names for them. Drawing on the mythology of the Roman god Jupiter (Greek Zeus), he used these four bodies to commemorate four of the god's lovers—three women and a boy. Io is the innermost of the four satellites, succeeded by Europa, Ganymede, and Callisto.

For nearly half a century, the only satellites known in the solar system were the Earth's Moon and the four galilean satellites of Jupiter. Then in 1655 the Dutch astronomer Christian Huygens discovered a satellite of Saturn that he named Titan, again drawing on the Greco-Roman pantheon. A few years later the French observer Jean Cassini found four more satellites of Saturn, equaling

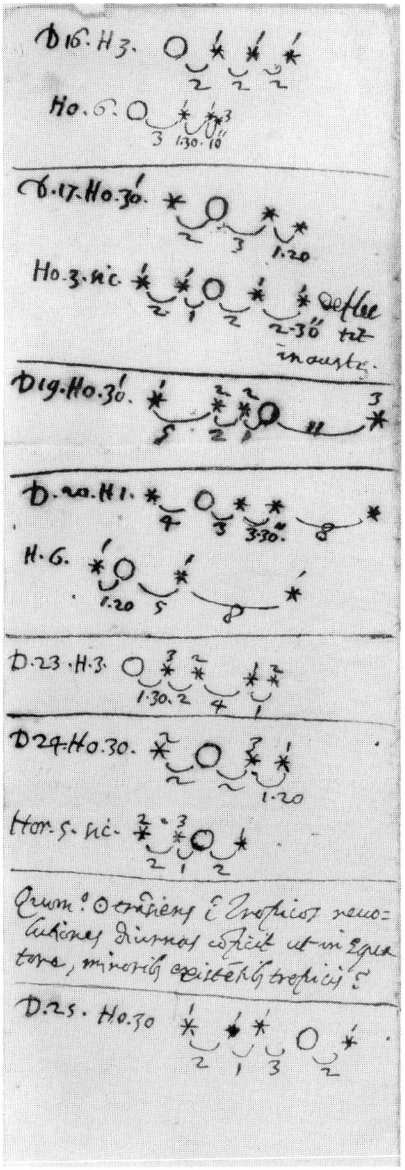

In 1610 Galileo discovered the satellites that now bear his name. This page from his observing notebook shows the satellites as star-like symbols that move back and forth with respect to the planet—motion he recognized as that of objects orbiting in the same plane, viewed edge-on.

the discovery record set by Galileo. William Herschel, the discoverer of Uranus, also identified four satellites, two each of Uranus (in 1787) and Saturn (in 1789). By the end of the eighteenth century, 14 satellites were known: one for Earth, two for Uranus, four for Jupiter, and seven for Saturn.

Telescopic discoveries continued in the nineteenth and twentieth centuries. Seth Nicholson of Mt. Wilson Observatory joined the select group of most productive discoverers with four small satellites of Jupiter, found between 1914 and 1951. (Only two satellites were discovered in the inner solar system, however: the small martian moons Phobos and Deimos; satellites are largely a phenomenon of the giant planets.) Still more moons were found when the Voyager spacecraft flew past each of the giant planets in turn. By 1990 the list contained 61 names, most small objects less than 100 km across.

The satellites are not the only small objects in the outer solar system. Another of Galileo's early telescopic discoveries was of two blurry "companions" of Saturn, which Christian Huygens later showed to be broad, flat rings surrounding the planet, composed of billions of tiny satellites orbiting in the same plane. For a long time Saturn was thought to be the only planet with rings. In the 1970s and 1980s, however, fainter ring systems were discovered girding Jupiter, Uranus, and Neptune. The rings that once seemed peculiar to Saturn are now viewed as a basic aspect of the giant planets, although many features of planetary rings remain mysterious.

Finally, two more objects should be discussed in a chapter on small bodies in the outer solar system: Pluto and its satellite Charon. Although a legitimate planet in terms of its orbit, Pluto is unlike the giants of the outer solar system. With less than a quarter the mass of our own Moon, Pluto is by far the smallest of the planets, and it actually ranks below half a dozen of the satellites in size. As a small, icy body far from the Sun, Pluto most closely resembles Neptune's satellite Triton.

The Jupiter System

Befitting its role as king of the planets, Jupiter boasts one of the largest satellite systems, with 16 known companions. Besides the four

The Satellites

Name	Diameter (km)	Period (Days)*	Discovery
EARTH			
Moon	3476	27.32	
MARS			
Phobos	23	0.32	1877
Deimos	13	1.26	1877
JUPITER			
Metis	20	0.29	1979
Adrastea	40	0.30	1979
Amalthea	200	0.50	1892
Thebe	90	0.67	1979
Io	3630	1.77	1610
Europa	3138	3.55	1610
Ganymede	5262	7.16	1610
Callisto	4800	16.7	1610
Leda	15	239	1974
Himalia	180	251	1904
Lysithea	40	259	1938
Elara	80	260	1905
Ananke	30	631R	1951
Carme	40	692R	1938
Pasiphae	40	735R	1908
Sinope	40	758R	1914
SATURN			
Pan	20	0.58	1990
Atlas	40	0.60	1980
Prometheus	80	0.61	1980
Pandora	100	0.63	1980
Janus	190	0.69	1966
Epimetheus	120	0.69	1980
Mimas	394	0.94	1789
Enceladus	502	1.37	1789
Tethys	1048	1.89	1684
Telesto	25	1.89	1980
Calypso	25	1.89	1980
Dione	1120	2.74	1684
Helene	30	2.74	1980
Rhea	1530	4.52	1672
Titan	5150	15.9	1655
Hyperion	270	21.3	1848
Iapetus	1435	79.3	1671
Phoebe	220	550R	1898
URANUS			
Cordelia	40	0.34	1986
Ophelia	50	0.38	1986
Bianca	50	0.44	1986
Cressida	60	0.46	1986
Desdemona	60	0.48	1986
Juliet	80	0.50	1986
Portia	80	0.51	1986
Rosalind	60	0.56	1986
Belinda	60	0.63	1986
Puck	170	0.76	1985
Miranda	485	1.41	1948
Ariel	1160	2.52	1851
Umbriel	1190	4.14	1851
Titania	1610	8.71	1787
Oberon	1550	13.5	1787
NEPTUNE			
Naiad	50	0.30	1989
Thalassa	90	0.31	1989
Despina	150	0.33	1989
Galatea	150	0.40	1989
Larissa	200	0.55	1989
Proteus	400	1.12	1989
Triton	2720	5.88R	1846
Nereid	340	360	1949
PLUTO			
Charon	1200	6.39	1978

**R indicates retrograde (here, opposite to direction of planet)*

large satellites discovered in 1610 by Galileo, the remaining 12, all of them small, conveniently divide themselves into three groups of four each.

Four lie close to the planet, well inside the orbit of Io. One of these, Amalthea, has been known for about a century, while the other three were discovered by Voyager in 1979. Closely associated with the innermost two moons are the tenuous rings of Jupiter, also discovered by Voyager—mere smoke rings of tiny particles, presumably eroded from the surface of these two satellites by the constant bombardment of micrometeorites and charged particles in the jovian magnetosphere.

Eight other moons lie far outside the system of the galilean satellites. Four are in highly inclined orbits (orbits not in the plane of the planet's equator); beyond these are four that circle the planet in backward, or retrograde, direction. All of these outer satellites are thought to be captured asteroids. Perhaps the two groups were formed in collisions and are fragments from the two original parent bodies.

Most of the interest in the jovian system focuses on the galilean satellites, which occupy evenly spaced orbits at distances from a half-

The Largest Satellites

Name	Diameter (km)	Mass (Moon = 1)	Density (Water = 1)	Surface Composition	Atmosphere (Pressure in Bars)
Ganymede	5262	2.0	1.9	dirty H_2O ice	none
Titan	5150	1.9	1.9	unknown	1.6
Callisto	4800	1.5	1.8	dirty H_2O ice	none
Io	3630	1.2	3.5	SO_2 ice, sulfur	none
Moon	3476	1.0	3.3	igneous silicates	none
Europa	3138	0.7	3.0	H_2O ice	none
Triton	2720	0.3	2.1	N_2 ice	0.00001

million kilometers for Io (about the same distance as the Moon from Earth) to nearly two million kilometers for Callisto. Io and Europa are about the size of our Moon, while Callisto and Ganymede — the largest satellite in the planetary system — are similar in size (but not in mass) to the planet Mercury. If they were orbiting the Sun instead of Jupiter, we would be pleased to call them planets and might by now have sent individual space missions to investigate them. Fortunately for the planetary scientist, however, they are close to Jupiter and therefore convenient to study as a group.

The smaller two galilean satellites, Io and Europa, have the highest density — about the same as our Moon, which is composed primarily of silicate rock. In contrast, the larger, outer satellites Ganymede and Callisto have lower densities, indicating that they are composed in part of water ice. In this respect the galilean satellites are similar to the planets, with the relatively small, dense objects close to the center and larger, less dense objects farther out. What does this tell us about the way planetary systems are constructed?

The planets close to the Sun are smaller and denser because temperatures where they formed did not favor the retention of ices like H_2O and light gases like hydrogen and helium. Could an analogous process have acted within the jovian satellite system? Did Io and Europa form in a region too warm for water to condense, while Ganymede and Callisto were cooler and hence able to incorporate water? Recall the birth of Jupiter. Because this planet is so large, a great deal of energy was released as it attracted infalling material from the surrounding nebula. The remnant of that primordial heat gives Jupiter its internal energy source today. Thus it is entirely plausible that a difference in distance from a young, hot Jupiter should account for the present break in size and composition between the inner and outer galilean satellites.

Ganymede and Callisto are the first two objects we have encountered in this book that are substantially composed of water ice. What does the presence of ice mean to their internal structure and surface properties? One clue is the relatively low melting temperature of ice. While it is cold enough today for these satellites to be frozen to the core, only a modest amount of energy would be required to make the ice lose its strength and become slushy.

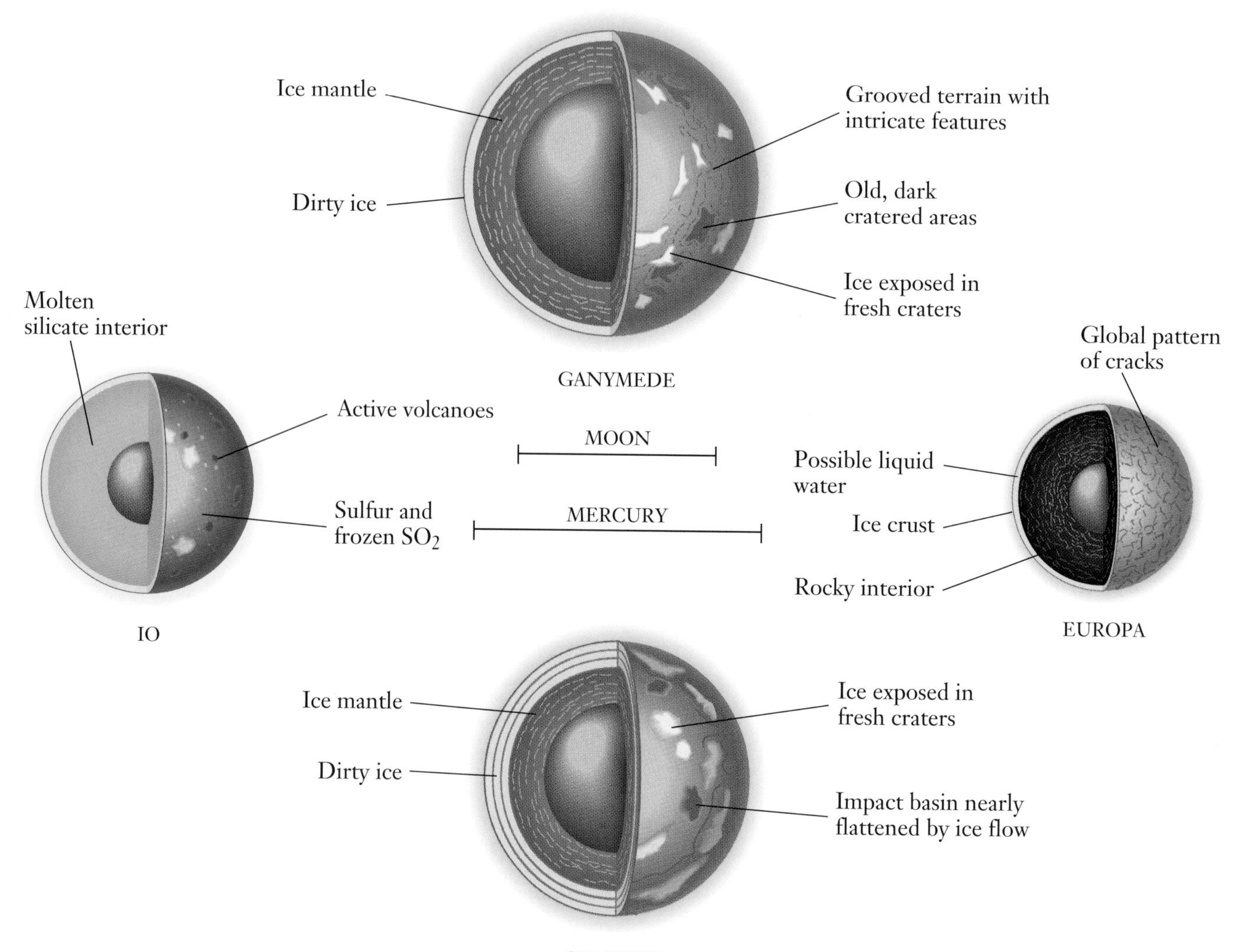

The inner and outer galilean satellites have strikingly different structures. Io and Europa are primarily rocky objects like our Moon, although Europa has an additional crust of ice about 50 km thick. The cores of Ganymede and Callisto are similar to those of Io and Europa, but with the addition of comparable masses of ice to form extensive mantles and crusts of lighter material.

Almost certainly temperatures at the time of formation of these satellites were high enough for the ice to melt. In that case, the rock and metal in their interiors would quickly sink to the center, while the slushy ice and water would rise to the top – a form of planetary differentiation similar to what we have seen in the terrestrial planets. The result is an object with a rocky or muddy core surrounded by a mantle and crust of ice. We can think of Callisto and Ganymede as silicate cores the size of Io or Europa encased in similar masses of ice.

What kind of geology characterizes an object nearly as large as Mars with a mantle and crust of ice? This is a question that tantalized planetary geologists anticipating the first Voyager encounter with Jupiter. The spacecraft images of Callisto and Ganymede provide some interesting answers.

Callisto is a good place to start because its geology is relatively simple. Heavily cratered, like the Moon or Mercury, Callisto has evidently not experienced much internal activity, so it has been able to accumulate impact craters for a long time. At first glance, then, it looks just like any other heavily cratered planetary surface. Upon close inspection, however, the craters of Callisto are seen to be much more shallow than those of a planet like the Moon. The explanation lies in the properties of ice, which deforms more easily than rock,

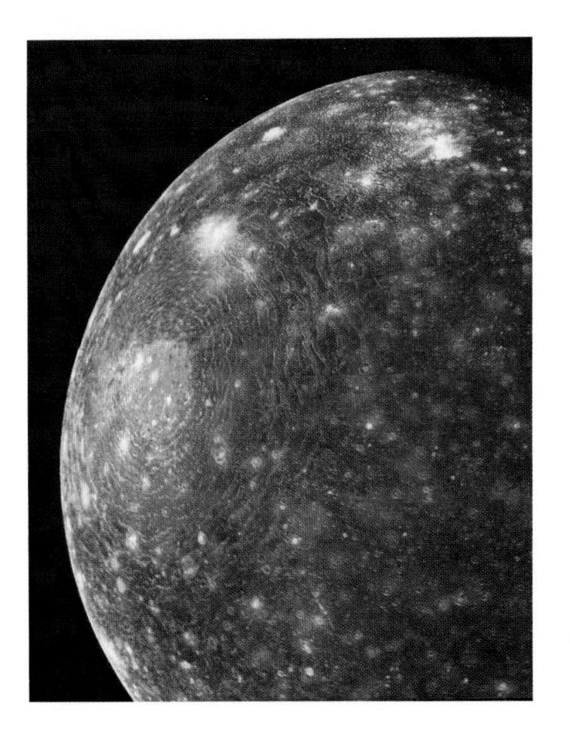

Callisto has the most heavily cratered surface, and the least evidence of internal activity, among the galilean satellites. The craters and impact basins on its surface, however, have been modified by deformation of the slightly plastic ice, which does not preserve topographic structure as well as the rocky surfaces of the terrestrial planets.

gradually slumping to fill in the crater floors and reduce altitude differences across the surface.

Ganymede is more interesting but also more challenging. About half of its surface consists of ancient cratered areas like the surface of Callisto, but the other half has been modified by internal forces. Evidently Ganymede was at some time warm enough for its ice to become slushy and for eruptions of water to flood its surface and erase the old craters. Some areas are characterized by ice mountains in the form of long, parallel ridges on about the same scale as the Appalachian Mountains of the eastern United States. There is even a hint of crustal movement on Ganymede that seems analogous to plate tectonics on the Earth. In this respect, icy Ganymede may have

Ganymede is the largest satellite in the solar system. Much of the surface is heavily cratered like Callisto, but other parts show evidence of internal modification. Recent craters have bright halos of ejected fresh ice.

Europa is geologically unique, with its smooth icy surface crossed by a network of fine lines. The ice crust may float on a global ocean of water. A prime objective of the Galileo spacecraft is to better understand the geology of this strange world.

more in common with the Earth than do our closer relatives, Venus and Mars. Scientists await with great expectation the more detailed views of Ganymede to be returned beginning in 1996 by the Galileo spacecraft. With luck, some of these images will show features as small as 10 m across — 100 times better than the resolution of the best Voyager images.

Turning to the inner galilean satellites, we find quite different bodies. Europa, although composed mostly of silicate rock, has a surface coating of ice, possibly floating on a deep global ocean of water. If Europa really has a liquid ocean, it is the only solid object in the solar system besides Earth possessing liquid water. Unfortunately Europa was not photographed at close range during the two Voyager encounters in 1979, so we don't know much about its ice crust. Obtaining better coverage of Europa is one of the primary objectives of the Galileo mission.

Io is the strangest of the galilean satellites. Voyager scientists had expected this rocky object to look much like our Moon, but it turned out to be completely different, with a colorful surface dominated by active volcanic eruptions. We will return to this unique satellite after a look at the satellite and ring systems of the other giant planets.

The Saturn System

Saturn has the largest number of known satellites (18) and the biggest ring system, but only one of its satellites (Titan) is as large as the four galilean satellites of Jupiter. Titan has an extensive atmosphere, which makes it one of the most interesting of all the satellites; we will discuss it in greater detail later in this chapter.

Eighteen satellites orbit Saturn in a direct sense, approximately in

The arrangement of the rings of Saturn is shown in this illustration. Almost all of the mass is in the A, B, and C Rings; the E Ring, in particular, is extremely tenuous.

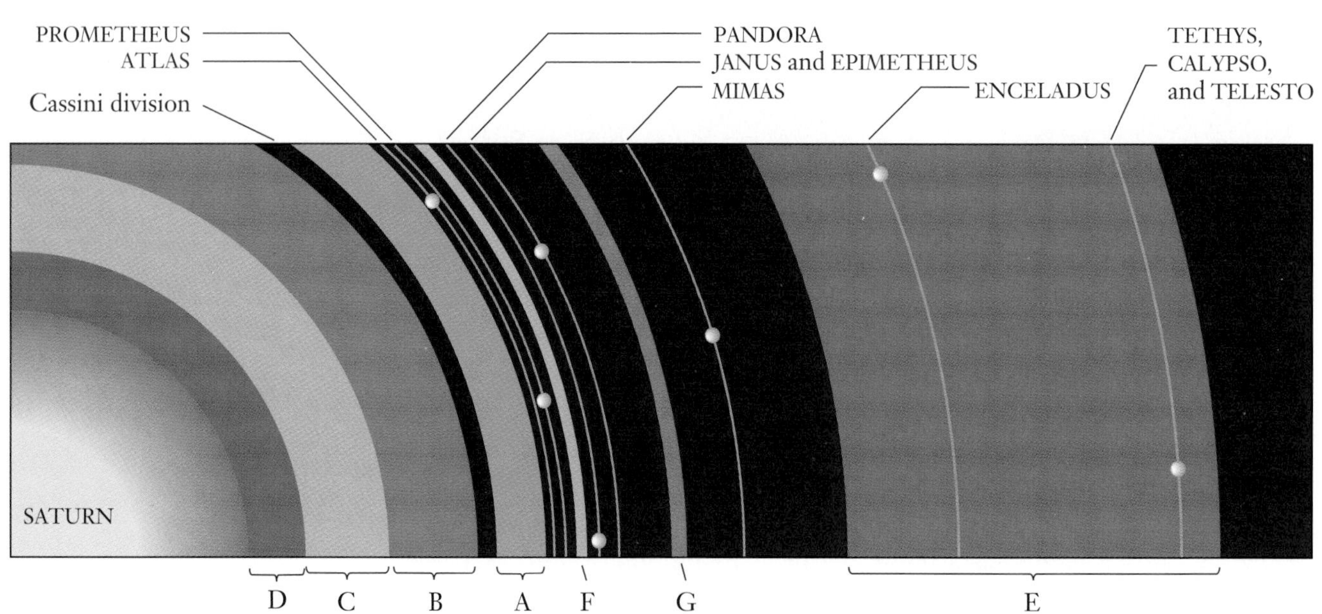

the plane of the rings. Only one small outer satellite, Phoebe, is a captured asteroid like the eight outer satellites of Jupiter. The regularity of the satellite orbits, together with the presence of broad, bright rings, gives Saturn the most majestic arrangement in the solar system.

In the case of Jupiter, the satellites are either large (the galileans) or small (less than 200 km in diameter). Saturn, in contrast, has a handful of icy satellites of intermediate size, typically about 1000 km in diameter. These are named (in order outward from the planet) Mimas, Enceladus, Tethys, Dione, Rhea, and Iapetus. These saturnian satellites have the lowest densities, and hence the largest proportion of water ice, of any of the measured satellites in the solar system. Perhaps the presence of such objects is related to the existence of the extensive rings, themselves made up of myriads of bright icy particles. There is no progression of increasing size and decreasing density as there was in the case of the galilean satellites, suggesting that Saturn was never warm enough to purge water ice from its inner satellites.

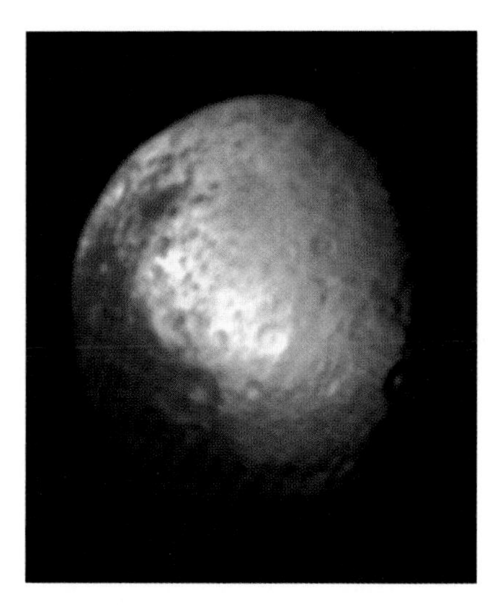

Saturn's satellite Iapetus has two faces; one hemisphere is composed of bright white ice, while the other is as dark as asphalt paving. We do not understand the origin of this dark carbonaceous material, nor why it coats just one side of the satellite.

Geologically, each of the icy satellites of Saturn has its own peculiar history. All are cratered, but here there is no slumping or flow of ice to soften the craters; at the distance of Saturn from the Sun, the ice is so solidly frozen that its properties are similar to those of rock. When we look down at the heavily cratered surface of Rhea, for instance, it looks very similar to the lunar highlands, even though we are seeing bright white ice rather than dark grey rock.

In addition to their craters, the satellites of Saturn exhibit varying degrees of internal geologic activity. Some, like Mimas, seem dead worlds. At the opposite extreme, little Enceladus exposes fresh surface areas that indicate continuing surface modification, presumably by release of water in a kind of "water volcanism." We do not know why Mimas is dead and Enceladus is alive, nor are many other aspects of these different satellites clearly understood. The next opportunity to learn more about these worlds will not come before 2004, when the planned Cassini mission to Saturn is scheduled to arrive.

Saturn is one of the most beautiful sights that can be seen with a telescope, and more than one professional astronomer has been

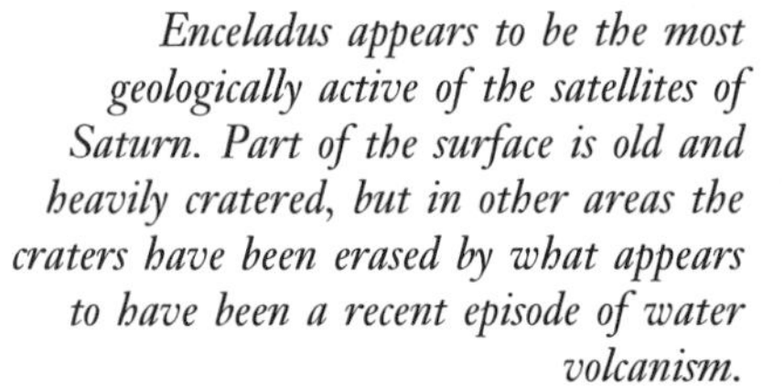

Enceladus appears to be the most geologically active of the satellites of Saturn. Part of the surface is old and heavily cratered, but in other areas the craters have been erased by what appears to have been a recent episode of water volcanism.

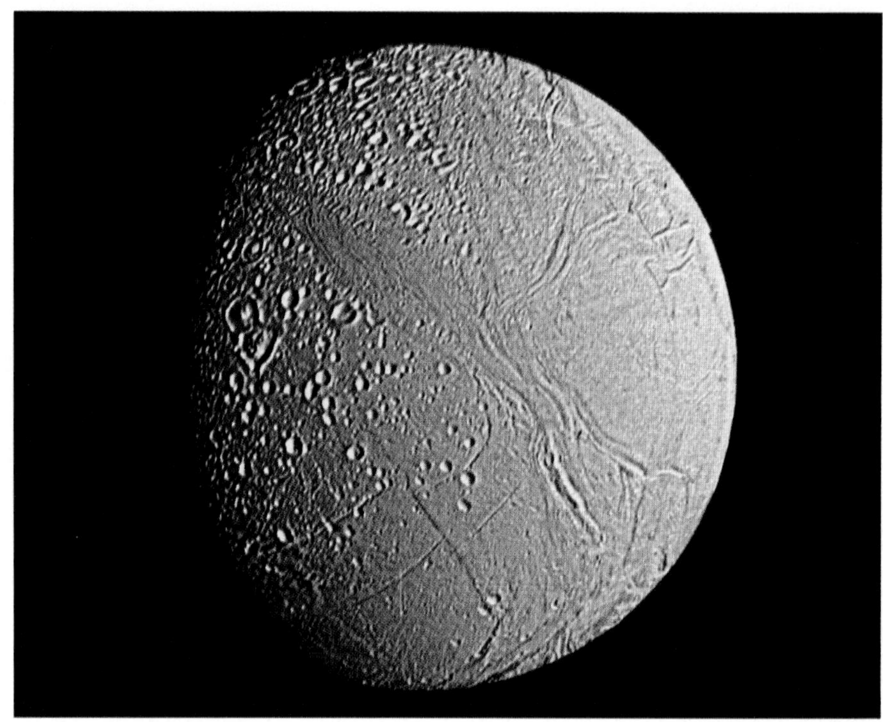

inspired to enter this field by childhood views of Saturn through an amateur telescope. The planet's most magnificent attribute is its system of rings. These surround the planet in the plane of its equator, which is titled by 27 degrees with respect to the common plane of the solar system. Thus we see the rings from many aspects over the course of Saturn's 30-year trip around the Sun, and they present a fascinating drama as they change year by year from fully open to edge on, at which time they may briefly disappear altogether.

The rings are invisible when seen edge on because they are fantastically thin. The width of the main rings is 70,000 km, yet their thickness is only about 20 m. If we made a scale model of the rings out of paper the thickness of the sheets in this book, we would have to make the rings 1 km across – about 8 city blocks. On this scale, Saturn itself would loom as high as an 80-story building.

The particles that make up the rings are composed primarily of water ice, like the satellites, and range in size from sand grains to

The dark underside of the rings of Saturn is shown in this Voyager photograph, obtained from a vantage point never accessible to Earth-based astronomers. The outer A Ring and inner C Ring are both translucent, but no sunlight penetrates the B Ring in the center.

house-sized boulders. A view from inside the rings would probably resemble a bright cloud of floating snowflakes and hailstones, including a number of snowballs and larger aggregates of smaller particles.

The main rings of Saturn are broad and flat, with only a handful of gaps. However, spacecraft images have revealed a few very slender rings, one (called the F Ring) outside the span of the main rings and several embedded in the gaps within them. As we will discuss later, there are many mysteries associated with the structure of planetary rings. But first let us examine the satellites and ring systems of the last two giant planets.

The Uranus and Neptune Systems

Not until the Voyager spacecraft arrived in the 1980s did scientists begin to appreciate the extent and complexity of the satellite and ring systems of Uranus and Neptune. Uranus has 15 known satellites while Neptune has 8, but only Neptune's large satellite Triton is in the same size class with the galilean satellites of Jupiter. Most of the satellites of these two planets are small, dark objects of apparently primitive composition – objects that presumably date back to the formative period of the solar system and have not been much modified since.

The rings and satellites of Uranus are tilted, just like the planet. All the satellites of Uranus have regular, essentially circular orbits. In contrast, Neptune's largest satellite, Triton, circles the planet in a retrograde direction, while its outermost satellite, Nereid, moves in an orbit of great eccentricity. Dynamically, these two Neptune satellites are among the most peculiar objects in the solar system.

Although Uranus has no large satellite, it does possess five objects of intermediate size (up to 1000 km in diameter). Like the intermediate satellites of Saturn, these small worlds have complex and varied geologic histories. In addition to the ubiquitous impact craters, they have experienced low-temperature volcanism, or cryo-volcanism. Since temperatures this far from the Sun are too low for rock or even ordinary water ice to melt, scientists have invoked more exotic lavas, such as mixtures of ammonia and water.

We have only limited knowledge of the composition of the satellites of Uranus and Neptune. The five intermediate Uranus satellites all have water ice on their surfaces, but they are darker than

The appearance of the rings of Uranus depends strongly on the direction from which we observe them. The first image shows the narrow, dark rings that constitute most of the mass in the ring system. However, the second image, taken looking toward the Sun, reveals many additional fine particles of dust, which are illuminated like dust on your car windshield when you are driving into the Sun.

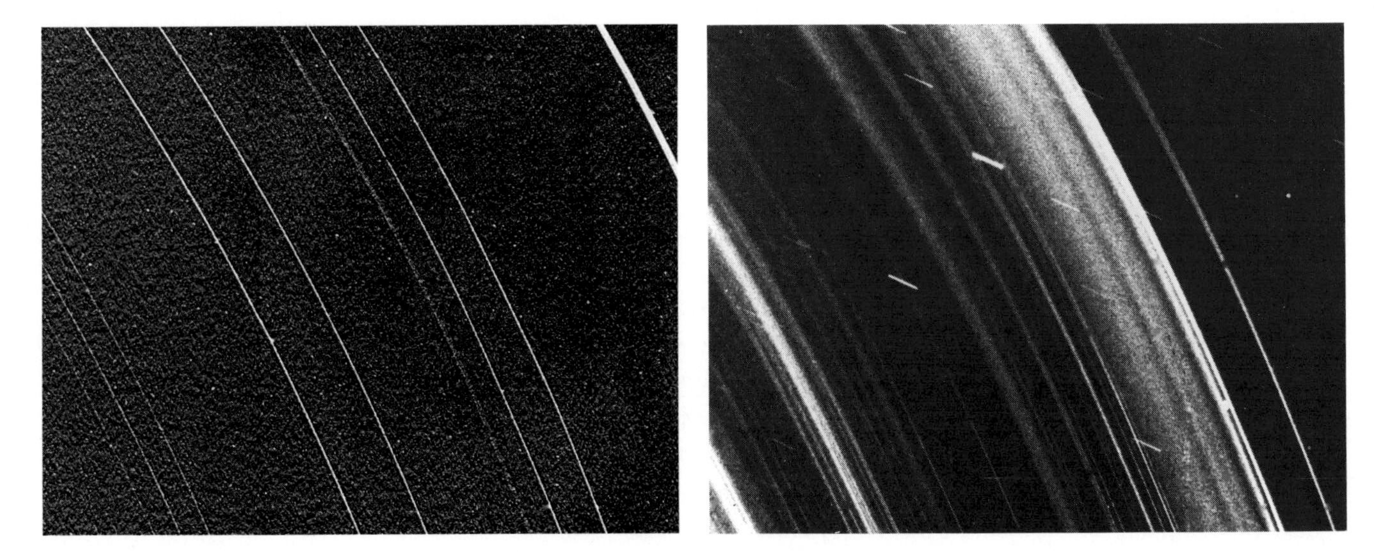

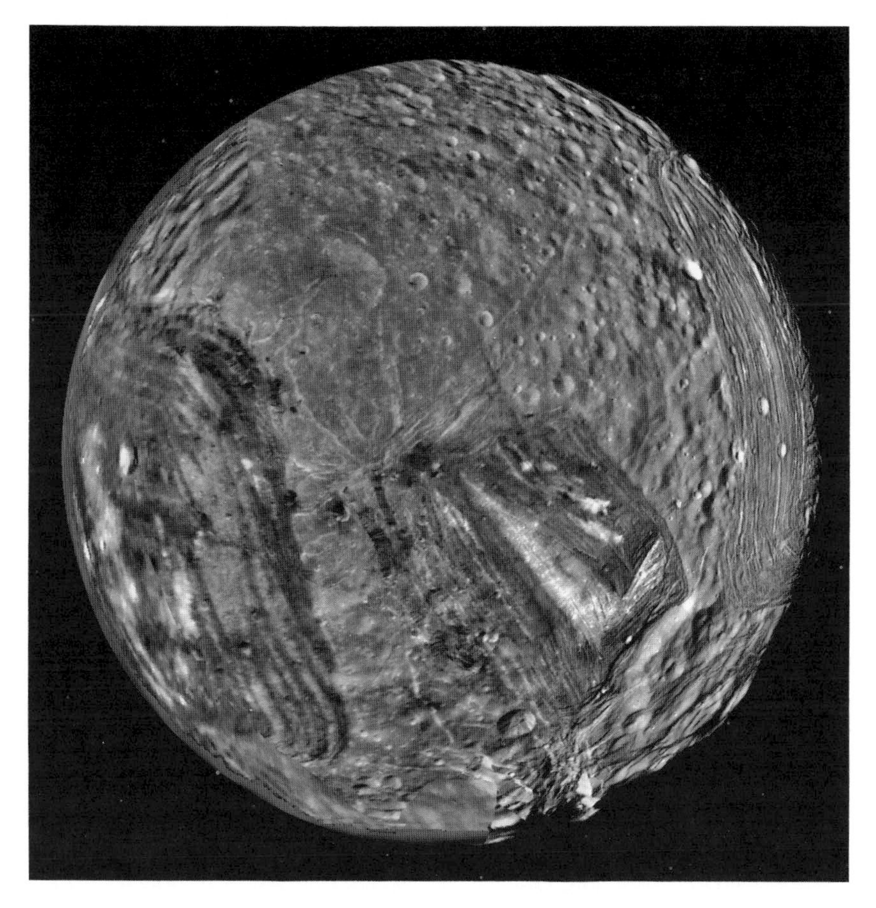

Miranda is the most geologically complex of the satellites of Uranus. Its surface shows areas of intense deformation suggestive of large-scale overturn of the surface, but there is no satisfactory explanation for the events that precipitated this situation.

the bright, icy satellites of Saturn. Presumably there is more dirt mixed with the ice. For the satellites large enough to have their densities measured, we find a bulk composition that includes more rock and less ice than the satellites of Saturn. This result contradicts the simple expectation that the farther we go from the Sun, the greater proportion of ice there should be. These satellites must be telling us something important about the conditions under which they formed, but we cannot yet decipher the message.

The rings of Uranus and Neptune are strikingly different from those of Saturn. Most obviously, they are less massive, by factors of nearly 1 million and 1 billion, respectively. Second, the ring particles are black, almost as dark as lumps of coal, instead of the bright white ice of the Saturn rings. Finally, the rings of Uranus and Neptune are slender and narrow with large open gaps between them, not broad and flat like the main rings of Saturn. Rather, they resemble the handful of narrow rings of Saturn discovered by the Pioneer and Voyager spacecraft.

Understanding these narrow rings has proved difficult. A multitude of small, orbiting particles is expected to spread gradually and finally dissipate. The main rings of Saturn have spread in this way, although they have been confined by some gravitational mechanism, since they have not dissipated. Stronger forces are required to confine the narrow rings and keep their particles moving together.

The rings of Neptune are even less substantial than those of Uranus, and to photograph them the Voyager cameras had to look back toward the Sun. The brightest outer ring is characterized by clumping or condensation of the particles in some regions, while other parts of the ring are much more tenuous.

Ring Mysteries

All four of the giant planets have rings, but these are very different in character. Among them, however, we can distinguish several basic ring types. Some very faint rings, for example the rings of Jupiter, are not confined to a plane at all, and these seem to be little more than unstructured, tenuous clouds of very small particles. Of primary interest are the broad rings, the narrow rings, and the ring gaps.

The broad, flat rings of Saturn are called by astronomers the A, B, and C Rings. With the exception of a few discrete gaps, these are nearly continuous, thin sheets of small orbiting particles. As revealed by the Voyager cameras, however, these broad rings are not without structure. Seen at close range, they break up into thousands of concentric patterns of alternating light and dark, indicating variations in the numbers of ring particles. Looked at even more closely, many of these patterns turn out to be spirals, like the grooves in an old-fashioned phonograph record.

Most of this structure can be interpreted as waves in the surface of the rings, induced by the gravitational pull of the planet's

There is a great deal of intricate structure in the rings of Saturn, as indicated in this Voyager photograph of the dense B Ring. The structure shown here has scales of tens to hundreds of kilometers.

Planetary Rings

Planet	Ring Radius (km)	Ring Mass (Tons)	Ring Particles
Jupiter	128,000	10^7	fine dust
Saturn	140,000	10^{16}	bright; water ice
Uranus	51,000	10^{11}	dark; carbonaceous
Neptune	63,000	10^9	dark; carbonaceous

satellites. The existence of such waves shows us that the individual ring particles do not orbit the planet independently; rather, each interacts gravitationally with its neighbors. Consequently, waves can be produced and propagated across the rings like the waves on the surface of the ocean. The small-scale structure of this pattern constantly changes, just as the surface of the ocean never looks exactly the same from one moment to the next. But the basic pattern is stable, related to the distances and orbital periods of the satellites that induce the waves. This mechanism accounts for most of the structure in the main rings of Saturn. The gravitational influence of the satellites may also play a part in defining the boundaries of these three rings and containing them from further spreading.

The second kind of structure is the ring gaps, typically about 1000 km wide. In the case of Saturn, most of the gaps can be related to the presence of small satellites (or exceptionally large ring particles) embedded within the rings themselves. The embedded satellite sweeps out the gap and defines sharp boundaries on its inner and outer edges. As it orbits the planet, moreover, it also induces waves in the edges of the gaps, like the wake left by a moving ship. One of these small satellites, called Pan, was identified in the Saturn system from its wake. Analysis of the wave pattern led to a prediction of the location and brightness of the satellite, and when scientists turned to the 10-year-old Voyager pictures, they succeeded in finding the moonlet.

The third major type of feature, and the most difficult to understand, is the narrow rings seen at Saturn, Uranus, and Neptune. These rings are typically between 10 and 100 km in width, and they may not be circular. One theory suggests that the self-gravitation of the ring material is sufficient to confine the particles to this narrow space. Alternatively, some of these narrow rings seem to be defined by small satellites (called shepherd satellites) that orbit on either side of them and keep the ring particles from straying beyond strict boundaries. Such shepherd satellites have been identified for both the F Ring of Saturn and the largest ring of Uranus, but it is hard to imagine enough such objects to account for all of the 20-odd narrow rings so far discovered. The situation is further complicated by the detailed structure in these narrow rings, which has not been satisfactorily explained by any theory.

The basic fact that rings exist and that their structure is controlled in large part by the gravitation of nearby satellites is clear. Important details remain to be explained, however. Similarly, there is uncertainty about ring origins. Most scientists think that the rings are made of the remnants of shattered small satellites, dating from the early history of the solar system. Others suggest, however, that the rings may be relatively young. Perhaps the Cassini mission can help explain these mysteries when it arrives at Saturn early in the next century, since the Saturn rings appear to include examples of every type of ring structure identified on any of the giant planets.

Two shepherd satellites, Pandora and Prometheus, are shown here on either side of the thin F Ring of Saturn. The F Ring is gravitationally confined by these two small satellites.

Volcanic Io

Let us turn now to a few of the most interesting individual large satellites. As we have noted, most of the small bodies of the outer solar system are composed in part of water ice. There is one striking exception, however: volcanic Io, the innermost of the four galilean satellites.

The discovery of active volcanism on Io was the most dramatic event of the Voyager flybys of Jupiter. Eight volcanoes were seen erupting when Voyager 1 passed in March 1979, and six of these were still active four months later when Voyager 2 passed. These eruptions consisted of graceful plumes that extended hundreds of kilometers into space.

The material erupted is not silicate lava or steam or carbon dioxide—all of which are vented by terrestrial volcanoes—but sulfur and sulfur dioxide. Both of these can build up to high pressure in the crust of Io and be ejected to tremendous heights. As the rising plume cools, the sulfur and sulfur dioxide recondense as solid particles, which fall back to the surface in gentle "snowfalls" that extend as much as 1000 km from the vent. The sulfur dioxide snow is white, while sulfur forms red and orange deposits. The surface of Io is slowly buried in these deposits, so it is no surprise that impact craters have not been seen on Io's surface.

Io displays other types of volcanic activity in addition to the spectacular plume eruptions. Images of its surface show numerous shield volcanoes and twisting lava flows hundreds of kilometers long. From their colors, these congealed lava flows are thought to be sulfur. Further volcanic activity is indicated by hot spots, surface areas that are hundreds of degrees warmer than their frigid surroundings. (Note that on Io, where the average daytime temperature is −140 C, even an area at 20 C, the surface temperature of the Earth, would qualify as a hot spot.) The largest of these hot spots is a "lava lake" of liquid sulfur 200 km in diameter near the Loki volcano. Telescopic observations show that this Loki hot spot has been active for at least 15 years, and it accounts for about half of the total volcanic energy released by Io.

How can Io maintain this remarkable level of volcanism, which exceeds that of much larger planets such as Earth and Venus? The

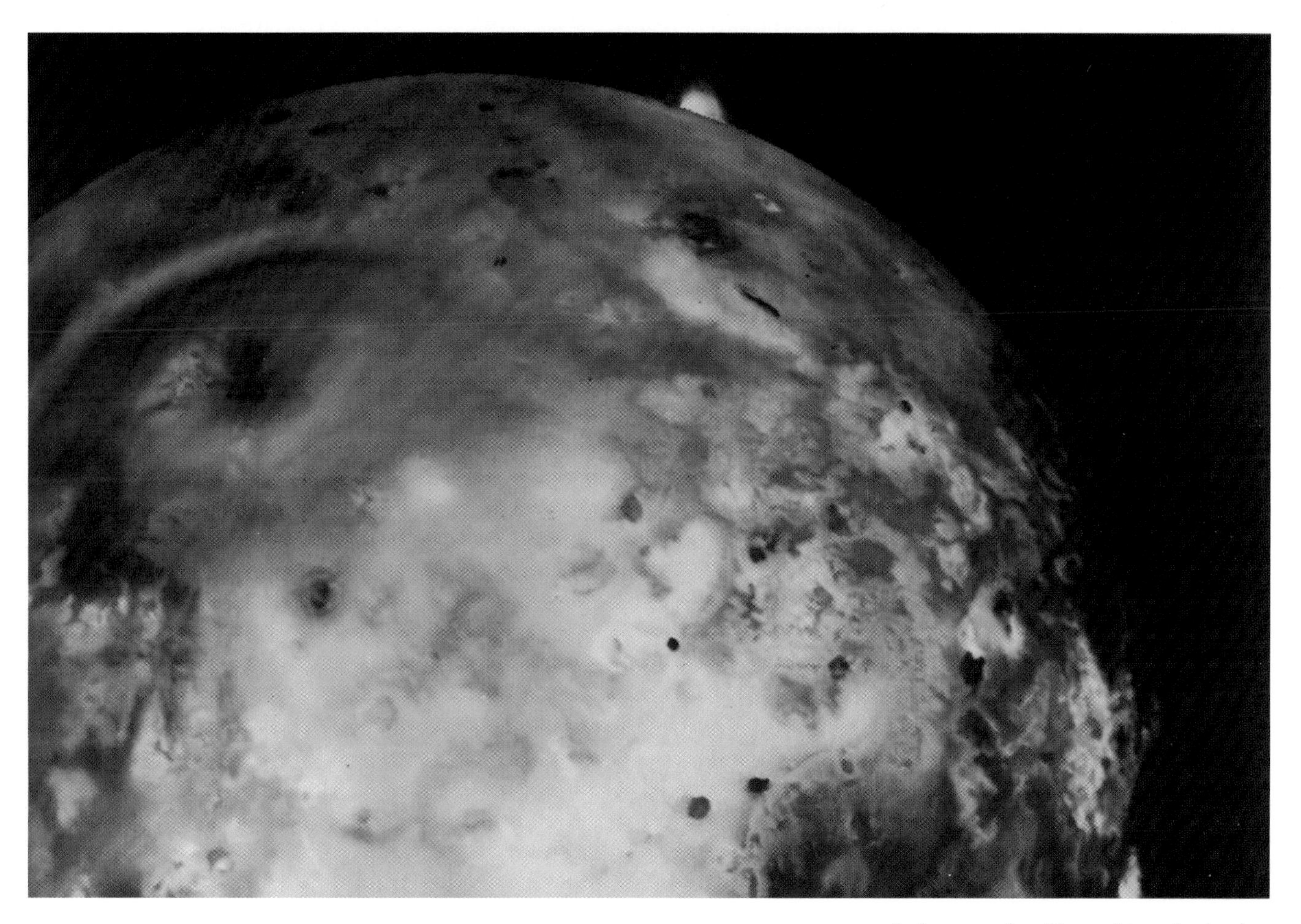

Io has a surface like nothing else in the solar system. Most of the material is sulfur and sulfur dioxide ejected from volcanic eruptions, such as the plume seen at the edge of this Voyager image.

answer lies in tidal heating of the satellite by Jupiter. Io is about the same distance from Jupiter as is our Moon from the Earth, yet Jupiter is more than 300 times more massive than Earth, causing tremendous tides on Io. These tides pull the satellite into an elongated shape, with a bulge several kilometers high extending toward Jupiter. If Io always kept exactly the same face turned toward Jupiter, this tidal bulge would not generate heat. However, Io's orbit is not exactly circular, because of gravitational perturbations from Europa and Ganymede. In its slightly eccentric orbit, Io twists back and forth with respect to Jupiter, at the same time moving nearer and farther from the planet on each revolution.

The most active volcanic area on Io is called Loki; it is centered near the dark "lava lake" in the lower right of this Voyager image. Telescopic data show that this area has been active ever since observations began in the mid-1970s. The colors are greatly exaggerated in this photograph, which has been processed to accentuate subtle detail.

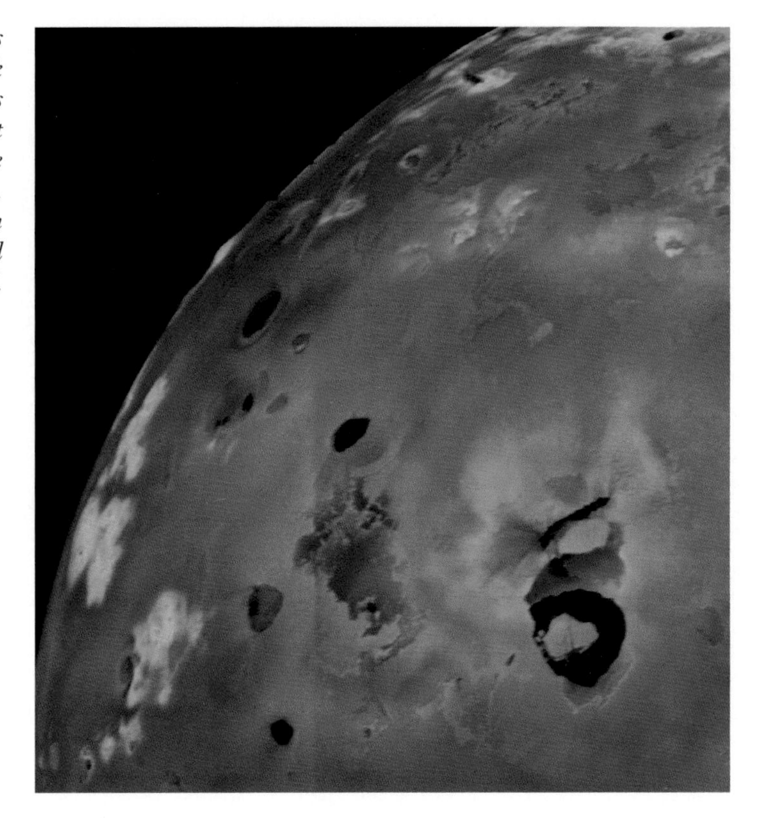

The twisting and flexing of the tidal bulge heats Io, much as repeated flexing of a wire coathanger heats the wire. Thus the complex interaction of orbit and tides pumps energy into Io, providing power to drive its volcanic eruptions. The inside is close to the melting point of rock, and the crust is constantly recycled by volcanic activity. Although Io was well mapped by Voyager, we expect that when re-imaged by the Galileo spacecraft after a 16-year interval, its surface will wear a partly unfamiliar face.

Cloudy Titan

Saturn's satellite Titan is unique among moons for the presence of a substantial atmosphere, discovered in 1944. Spectra initially showed absorptions due to methane gas, while subsequent observations established the presence of dense clouds, obscuring the surface from our view. Reflecting interest in this atmosphere, the Voyager 1 flyby of Saturn was designed to yield as much information as possible about Titan. Voyager passed within 4000 km and then flew behind the satellite (as seen from the Earth), so that its radio signal traversed successive paths through Titan's atmosphere, generating data from which scientists could reconstruct its atmospheric profile all the way down to the invisible surface. The measured surface pressure was 1.6 bars, higher than on any of the terrestrial planets except Venus.

Titan's atmosphere is primarily nitrogen, another respect in which Titan resembles the Earth. Methane and argon each amount to a few percent at most. Additional compounds detected spectroscopically in Titan's upper atmosphere include carbon monoxide, various hydrocarbons, and nitrogen compounds such as hydrogen cyanide. The discovery of cyanide was particularly interesting, since this molecule is the starting point for formation of some of the components of DNA, the fundamental genetic molecule essential to life on Earth.

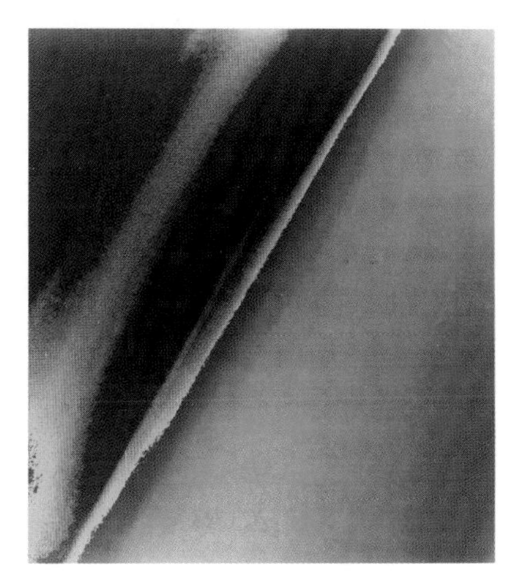

Haze layers can be seen far above the thick orange clouds of Titan in this high-resolution view from Voyager 1. The colors in this photograph are highly exaggerated.

The primary clouds on Titan are found in the bottom 10 km of the atmosphere; these are condensation clouds, probably composed of methane. Methane plays the same role in Titan's atmosphere as water does on Earth: the gas is only a minor constituent, but it condenses to form the major clouds. Much higher is a reddish haze or smog consisting of complex organic chemicals. Formed at an altitude of several hundred kilometers, this aerosol slowly settles downward, where it presumably has built up a deep layer of tarlike organic chemicals on the surface of Titan.

Titan's surface temperature is about −180 C, held uniform by the blanketing atmosphere. At such a low temperature, there may be seas of liquid methane and ethane. Organic compounds are chemically stable at Titan's temperatures, unlike the situation on the warmer, oxidizing Earth; so Titan's surface probably records a chemical history that goes back billions of years. Many people believe that this satellite will provide more insights into the early history of Earth's atmosphere, and even into the origin of life, than any other object in the solar system. For this reason, the planned Cassini mission to Saturn features an atmospheric probe to make direct measurements of the atmosphere and perhaps also the surface of this remarkable object.

Frigid Triton and Pluto

At the outer edges of the planetary system, 5 billion km from the Sun, lie Pluto and Triton. Pluto is the only planet not yet visited by a spacecraft. Our knowledge of its properties is limited, but it seems to be very similar to Neptune's largest satellite, Triton.

The origin of Pluto is a mystery, since this little world, with only a quarter the mass of our Moon, is completely unlike the giant planets. Because of its small size, Pluto was not discovered until 1930. Its orbit is the most elongated of any of the planets, varying in distance from the Sun between about 30 and 50 AU. It also has the most highly inclined orbit (17 degrees). These peculiarities have prompted the suggestion that Pluto is not a real planet at all. Perhaps astronomers will someday redesignate it as a very large comet or asteroid.

Much of the information we have on Pluto has been derived recently from the presence of its satellite Charon, discovered in 1978. The existence of Charon has allowed us to measure the mass and density of Pluto, revealing that it is about 80 percent rock and 20 percent ice. Additional observations made as Charon and Pluto have alternately passed in front of and behind each other (as seen from the Earth) have yielded crude maps of the surface of Pluto and demonstrated it contains frozen methane and nitrogen, while Charon seems to be composed of water ice. Finally, we have detected Pluto's tenuous atmosphere of nitrogen, currently at its maximum since Pluto just passed perihelion (the point in its orbit nearest the Sun) in 1989 and is therefore at maximum surface temperature. As Pluto recedes from the Sun over the next few decades, its atmosphere is expected to collapse and freeze onto the surface.

We know more about Triton, since it was observed at close range by the Voyager 2 spacecraft in 1989. The surface material of Triton is fresh ice or frost, with the very high average reflectivity of about 80 percent. This frost may include mixtures of water, methane, and nitrogen, all of which are frozen at Triton's low temperature of −236 C; at just 37 degrees above absolute zero, it is the coldest place in the solar system. Most potential atmospheric gas is frozen at these temperatures, but a small quantity of nitrogen vapor persists, just as on Pluto.

Voyager photos show that Triton's surface, like that of many other satellites in the outer solar system, reveals a long geological evolution. While there are some impact craters, many regions have also experienced cryo-volcanism. There are frozen "lava lakes" more than 100 km across and regions of jumbled or mountainous terrain that resemble the mountainous regions of Ganymede. A polar cap covers much of the southern hemisphere, apparently evaporating along its northern edge. This polar cap may consist of frozen

Triton, the large retrograde satellite of Neptune, was the final object visited by Voyager 2 in its grand tour. Triton turned out to be a remarkable object, with the coldest surface in the solar system and a complex and little understood geological history. The lighter areas toward the bottom of the image probably constitute an evaporating polar cap of frozen nitrogen, but no one is sure.

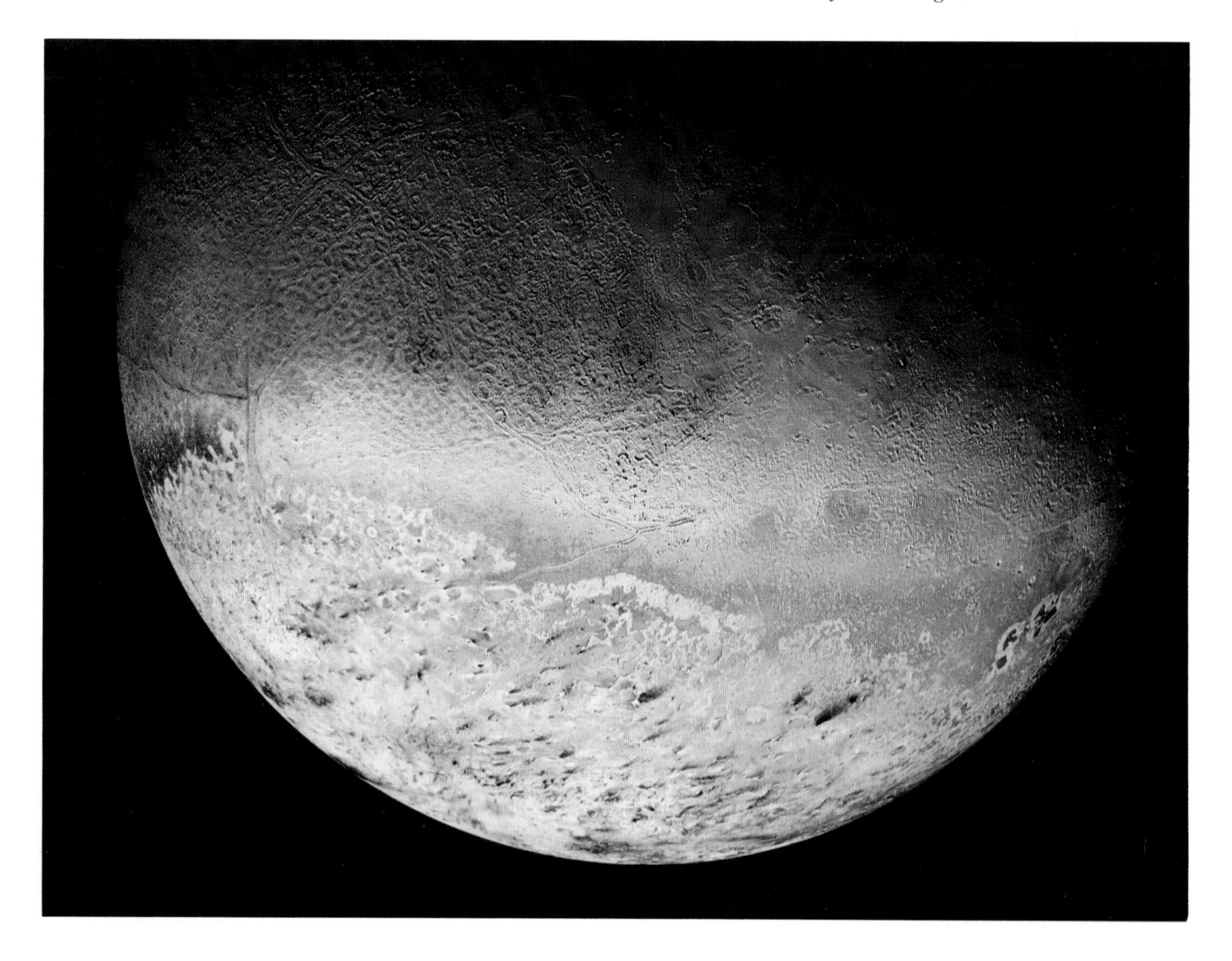

nitrogen, deposited during the previous winter. Remarkably, its evaporation seems to generate geysers or volcanic plumes of nitrogen that fountain to altitudes of about 10 km above the surface. These plumes differ from the volcanic plumes of Io in their composition and energy source, which is sunlight warming the surface rather than internal heat.

Triton has turned out to be a far more singular object than had been imagined before the Voyager encounter. Many planetary scientists would place it alongside Io and Titan as one of the most interesting satellites in the outer solar system. These discoveries also make us wonder what Pluto would look like if we could observe it up close. We now speculate that Pluto and Triton are similar; past experience suggests, however, that an actual close-up view of Pluto would yield many surprises and perhaps weaken this analogy. A flyby of Pluto remains a high priority in NASA's planetary exploration program.

Planet X?

What lies beyond Pluto? Billions of comets, certainly, as we will discuss in the next chapter. But is it possible that another planet lurks undiscovered in the deep freeze of the outer solar system? Have we really reached the limit of the planets? Those who hope there is a tenth planet yet to be discovered have often called this unknown world Planet X.

Many photographic surveys have been made, either as part of a direct search for Planet X or as a by-product of other efforts. Clyde Tombaugh, the discoverer of Pluto, is satisfied from such searches that no other planet exists that is even half as bright as Pluto. Planet X must be either very small, very dark, or very far away. The presence of another giant planet like Uranus or Neptune is clearly excluded. An icy object much smaller than Pluto could certainly be present, but we might not call such a body a planet. Several large asteroids (200 to 300 km in diameter) are known in the outer solar system, but these hardly count as full-scale planets. One of these objects, discovered in 1992, actually lies beyond the orbit of Pluto.

The search that led to the 1930 discovery of Pluto was motivated by reported irregularities in the motion of Uranus and Neptune, attributed to the gravitation of an unseen distant planet. However, Pluto is so small that it cannot have caused the reported problems with Uranus and Neptune. This leaves open the possibility of a massive but very dark object still undiscovered. The best way to search for a very dark object is not by reflected sunlight but by its emitted thermal radiation. In 1983 the Infrared Astronomy Satellite (IRAS) carried out an all-sky survey from Earth orbit that was much more sensitive to long-wave thermal radiation than any previous search. IRAS covered 99 percent of the sky before its detectors failed, and it found no dark planets. Unless by some extraordinary coincidence Planet X was in the 1 percent of the sky not observed, it seems that we can now dismiss this hypothesis. Most likely there is no planet, and the irregularities in Uranus and Neptune are not, and never were, real.

Pluto probably represents the edge of the planetary system. Indeed, in many ways the system ends with Neptune. Pluto is an anomaly, whatever its origin. Perhaps it is simply the largest surviving icy object of the type that formed the building blocks of the giant planets some 4.5 billion years ago. In a sense it marks, not the last of the planets, but the beginning of the realm of the comets.

REMNANTS OF CREATION

Comets, Asteroids, and the Origin of the Solar System

•••

Comet Halley is the best known of the left-over building blocks of rock and ice that remain in the solar system. In 1986, the European Giotto spacecraft was one of an international flotilla that studied the comet at close range.

We are all curious about origins: Where did we come from, and how did we become what we are today? A recurrent theme of this book is the evolution of the planets. Ironically, however, the planets themselves are largely mute on the conditions of their origin. Just as the study of an adult human population provides limited and ambiguous data on birth and the developments of early childhood, the mature planets retain only limited and ambiguous memories of the conditions of their formation and early history. Much more helpful would be examples of smaller, less modified objects — planetary building blocks — that have changed little since the formation of the solar system.

The key to the past lies in the smallest members of the solar system: comets and asteroids. Both the comets and the asteroids are remnants of creation, leftover material from the beginning of the solar system. Because of their small size, all comets and most asteroids have never been sufficiently heated to melt and differentiate. They are composed of relatively unmodified or primitive material that closely approximates the chemistry of the original nebula from which the solar system formed some 4.5 billion years ago.

Comets and asteroids are distinguished from each other by both their composition and location. The asteroids, mostly rocky objects, populate the wide gap between the orbits of Mars and Jupiter; probably they represent the raw materials available to form the terrestrial planets. The comets, in contrast, are derived from the outer parts of the solar system. Because formation temperatures were much lower in these regions, water and other ices were available to form objects composed in substantial part of such volatile compounds. Occasionally some of these icy bodies wander into the inner solar system, where they are heated by the Sun and develop transient atmospheres. When we see the wispy tail of escaping gas, we call this interloper a comet.

Although no spacecraft has landed on a comet or asteroid, we are fortunate to have samples of their material to study in our laboratories. Because they are small and have weak gravity, fragments of comets and asteroids produced by collisions easily escape and spread through the solar system, where they occasionally hit the Earth and are collected as meteorites or interplanetary dust.

Many astronomy books refer to the comets, asteroids, and meteorites as "debris." I prefer to think of them as artifacts from which we may be able to deduce events in the distant past. We study them in much the way an archeologist sifts through the ruins of past civilizations, in the hope that we too can locate a Rosetta Stone that will unlock some of the secrets of our origins.

Rocks from the Sky

Until fairly recently, the idea that extraterrestrial materials were reaching the surface of the Earth was scoffed at by educated persons, who placed stories of falling stones in the same category with tales of

Three important events occurred in Europe in 1492: Muslims and Jews were expelled from Spain, Columbus sailed to the New World, and a stony meteorite exploded and fell on the German town of Ensisheim, as illustrated in this contemporary woodcut. The meteorite is still on display in the town.

This 22-ton iron meteorite was discovered in Mexico in 1876, near the town of Bacubirito. Shown here is the crew of men that dug it out; it is exhibited today in a museum in Culiacan.

fairies and dragons. President Thomas Jefferson, a distinguished amateur scientist, reacted to claims of an 1807 meteorite fall in Connecticut by commenting that he could more easily believe that Yankee professors would lie than that stones would fall from the sky. His skepticism was perhaps appropriate, given the then-common opinion that meteorites were produced in the Earth's atmosphere by lightning, but it illustrates that rare events are easily dismissed. Within a few years of this statement, however, scientists had agreed that rocks do fall and that their origin is extraterrestrial.

Meteoritic materials are constantly reaching the Earth, and several falls are observed and recovered every year. The fragments that reach the ground are usually stones or metallic masses of only a kilogram or two, small enough to be held comfortably in your hand. Larger falls are produced when a mass of hundreds or thousands of kilograms strikes the atmosphere, often breaking up to scatter meteorites over many miles. Even larger projectiles strike at intervals of thousands of years, some of them exploding in the atmosphere while others produce impact craters when they crash into the surface.

The Antarctic continent has proved to be the most fertile hunting ground for meteorites; more than half of the known meteorites have been recovered from the ice, primarily by expeditions from the United States and Japan. The Antarctic ice cap does an excellent job of collecting and concentrating these rocks from the sky. Meteorites, encased in the ice, move with the glaciers that cover the continent. In regions of ablation this ice is exposed and stripped away by wind, leaving the meteorites on the surface, where they can be spotted and removed for later study. In addition to their large numbers, the Antarctic meteorites have the advantage of protection by the ice and cold from many forms of weathering and erosion that contaminate samples that fall in warmer and wetter climates.

Most meteorites on display in museums are composed of nearly pure metallic nickel-iron. Their extraterrestrial origin is obvious when we recall that iron and most other metals normally occur on

The primary source of meteorites today is the Antarctic continent. This stony meteorite is being photographed in place before being collected for return to the United States.

Earth in the form of oxides rather than as pure metals. Much more common, but less easily recognized for their origin, are a variety of stony meteorite types. Finally, there are rare examples of stony-iron meteorites, mixtures of stone and metal.

The most useful classifications of meteorites are based on their chemical history. If a meteorite is representative of the original materials out of which the solar system was made, we refer to it as a primitive meteorite, or chondrite (the term is derived from the presence in these meteorites of tiny spheres of rock called chondrules, presumably congealed droplets dating from the formation of these meteorites in the solar nebula). All primitive meteorites are stones. However, not all stones are primitive; some of the stony meteorites are fragments of asteroids that were once melted and differentiated, just like the terrestrial planets.

Radioactive dating reveals that all chondrites are close to 4.5 billion years old, a value associated with the age of the solar system. Since the majority of the meteorites that reach the Earth are chondrites, we can be confident that material still exists in the solar system that has remained relatively unchanged since before the planets were formed.

Although the surface of a meteorite is heated to incandescence during its brief plunge through the atmosphere, the heat pulse penetrates no more than a few centimeters into the interior, most of which remains cool and undisturbed. Even the outer layers have normally cooled by the time the meteorite strikes the ground, as shown by objects that have fallen on ice or snow without melting it. Thus primitive meteoritic material remains largely unaltered by its violent arrival.

The iron and stony-iron meteorites, as well as some types of stones, have experienced major chemical or physical change since their formation; these are called differentiated, or igneous, meteorites. These meteorites provide us with a window on the interiors of differentiated asteroids that have been shattered in collisions. Presumably the irons are fragments from their cores, the various igneous stones are samples of their mantles and crusts, and the rare stony-irons are derived from the core-mantle interface. While they can tell us many interesting things about the interiors of their

Many different kinds of stones fall from the sky, representing a wide variety of asteroid parent bodies. In the top row are the Allende carbonaceous meteorite (left), with white inclusions that may predate the solar nebula itself, and a fragment of the iron meteorite that was responsible for the formation of Meteor Crater in Arizona. The lower row shows the Imilac stony-iron pallasite (left), a beautiful mixture of green olivine crystals and metallic iron that originated deep within the mantle of a differentiated parent body, and the Mern primitive chondrite.

precursor bodies, such meteorites do not contribute directly to study of conditions at the time the solar system formed.

The most primitive meteorites are a special group called the carbonaceous meteorites. Relatively rich in carbon (a few percent by weight), as their name implies, they are also rich in volatile compounds like water, which combines chemically with other minerals to form clays. Carbonaceous meteorites are dark gray to black in color and are physically very weak; a fragment can generally be crushed between the fingers. Much of their carbon is in the form of complex organic compounds produced in the hydrogen-rich environment of the solar nebula, or perhaps still earlier in interstellar space. These organic materials include 16 different amino acids, 11 of which are rare on Earth. The most remarkable thing about these and other amino acids in meteorites is that they include equal numbers of left-handed and right-handed forms. Amino acids can occur or be synthesized having either of these kinds of symmetry, but life on Earth has evolved using, in fact, only the left-handed versions to make proteins. The presence of both symmetries in the meteorites demonstrates that no contamination has taken place since they arrived on Earth, and it suggests that these organic compounds formed in space without the intervention of living things.

Meteorite Parent Bodies

In two cases the parent bodies of meteorites are known. Several lunar rocks have been identified recently, as well as a larger number of objects that almost certainly originated on Mars. One of the lunar meteorites is privately owned, making it the only lunar sample not rigidly controlled by the governments of the United States, Russia, or Japan. It is not clear what the value of this moonrock might be on the open market, although presumably it is a very small fraction of the billions of dollars expended in the Apollo program, the source of most of the other lunar samples on the Earth.

Meteorites are fragments, testifying to the continual breakup and erosion of the larger bodies from which they originate. One useful approach to the study of meteorites is to use their properties to characterize these unknown parent bodies.

Meteorites fall from all directions, and it is usually not possible to determine their flight paths with sufficient accuracy to reconstruct their orbits in space before they encountered the Earth. Such reconstruction has been possible for three chondrites, however, based on precision photographs of their fiery entry into the atmosphere. All three were on elongated orbits that stretched from just inside the orbit of the Earth out to about 3 AU, in the middle of the asteroid belt between Mars and Jupiter—orbits pretty much what we would expect for fragments from asteroids.

Looking at the chondrites themselves, we conclude that they originated from relatively small bodies that formed directly from dust condensing out of the cooling solar nebula. Large parent bodies are not possible: they would have retained too much internal heat, generated by natural radioactivity, and this heat would have modified their chemistry in ways we do not observe. Calculations show that in order to have escaped unacceptable heating levels, the parent bodies must have been no more than a few hundred kilometers in diameter. The chemical and isotopic variety among the primitive meteorites indicate that many different parent bodies are represented in our meteorite collections.

The igneous meteorites are fragments of differentiated parent bodies. From detailed analysis, it is clear that several dozen distinct parent bodies were involved. We can set some limits on their size, since the iron meteorites can be analyzed to derive cooling rates and calculate maximum diameters of a few hundred kilometers. Thus the differentiated parent bodies were at least as small as the parent bodies of the chondrites, and a characteristic other than size was responsible for the fact that one group of objects heated and differentiated while the other remained in a primitive state. No one knows why one set of parent bodies differentiated and the other did not.

Thus we conclude that there were a minimum of several dozen parents for the meteorites now in our collections and that these parents had diverse chemical and heating histories. Clearly excluded is the old idea of an exploded planet as the source of the meteorites. Also unlikely is a source among the ice-rich comets. Probably many of these parent bodies persist today among the asteroids.

Asteroids

Most asteroids orbit the Sun at distances of 2.2 to 3.3 AU, in the region called the asteroid belt. This wide gap between the orbits of Mars and Jupiter represents one of the few locations in the solar system where such orbits are stable. Once an object finds itself in this region, it can survive for the lifetime of the solar system, barring catastrophic collisions with its neighbors.

Most asteroids orbit the Sun between Jupiter and Mars. Only within this asteroid belt, between 2.2 and 3.3 AU from the Sun, are orbits stable for the 4.5-billion year lifetime of the solar system. Individual orbits for five asteroids are shown: Ceres, Juno, and Pallas in the belt, and near-Earth asteroids Apollo and Icarus. Icarus may be a dead comet nucleus.

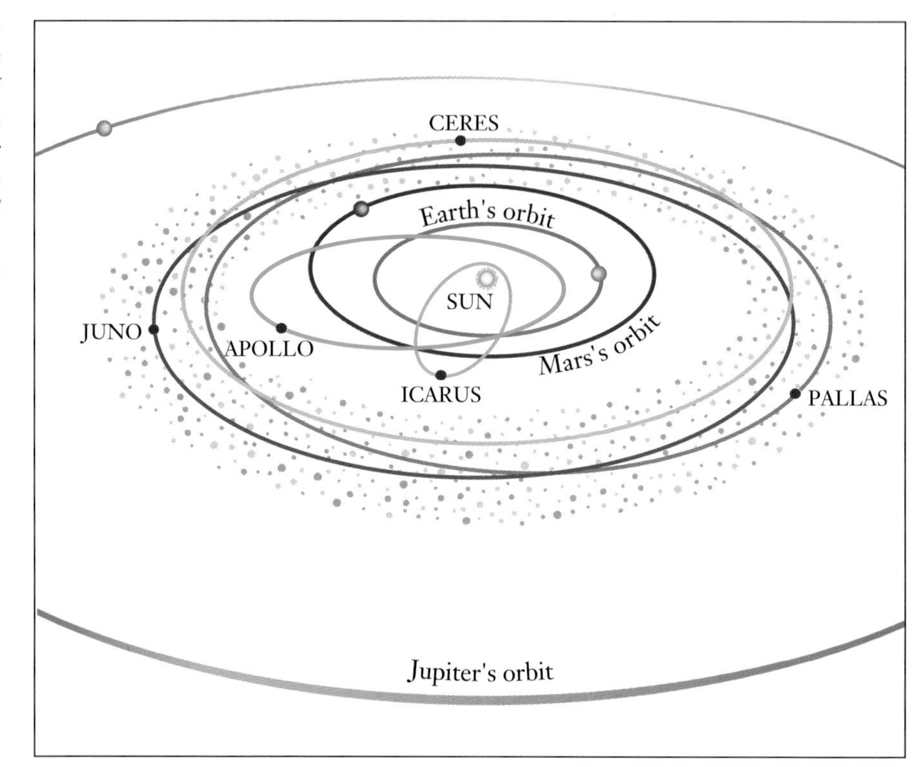

Ceres, the largest asteroid, was discovered in 1801. While it is just at the limit of naked-eye visibility, Ceres would never have been found without a telescopic search. Ceres, with a diameter of just under 1000 km, accounts for about a quarter of the total mass in the asteroid belt. The next-largest asteroids are about half as large. About 6000 asteroids have been discovered, and our census of the larger ones is now fairly complete. It is estimated that 99 percent of the objects 100 km or more in diameter are known, and discovery should be at least 50 percent complete for diameters down to 10 km. In contrast, we have found very few objects as small as 1 km in diameter, and there may be nearly a million down to this size threshold.

Telescopic studies reveal many properties of individual asteroids, such as their size, period of rotation, and surface reflectivity. Additional information can be obtained from the spectrum of reflected sunlight. Different minerals have different colors in both the

The Largest Asteroids

NAME	DISCOVERY	DIAMETER (KM)	TYPE
Ceres	1801	940	carbonaceous
Pallas	1802	540	carbonaceous
Vesta	1807	510	basaltic
Hygeia	1849	410	carbonaceous
Interamnia	1910	310	carbonaceous
Davida	1903	310	carbonaceous
Cybele	1861	280	carbonaceous
Europa	1868	280	carbonaceous
Sylvia	1866	275	carbonaceous
Juno	1804	265	stony
Psyche	1852	265	metallic

visible and infrared; therefore variations with wavelength in the reflectivity of its surface material can indicate the composition of the asteroid. While not as rigorously diagnostic as the sharp spectral lines produced when light passes through a gas, the broad absorption features in the spectrum of an asteroid are often sufficient to identify the major minerals present.

Most asteroids fall into one of two classes based on their reflectivity: they are either very dark (reflecting only 3 to 5 percent of incident sunlight) or moderately bright (15 to 25 percent reflectivity). A similar distinction exists in their spectra; the dark asteroids have spectra similar to the carbonaceous meteorites, while the lighter ones look like the most common types of chondrites. The dark asteroids are called C (carbonaceous) and the light are called S (stony). A third group, the M asteroids, is metallic, like the iron meteorites, while other rare asteroid classes appear to correspond to different types of igneous meteorites.

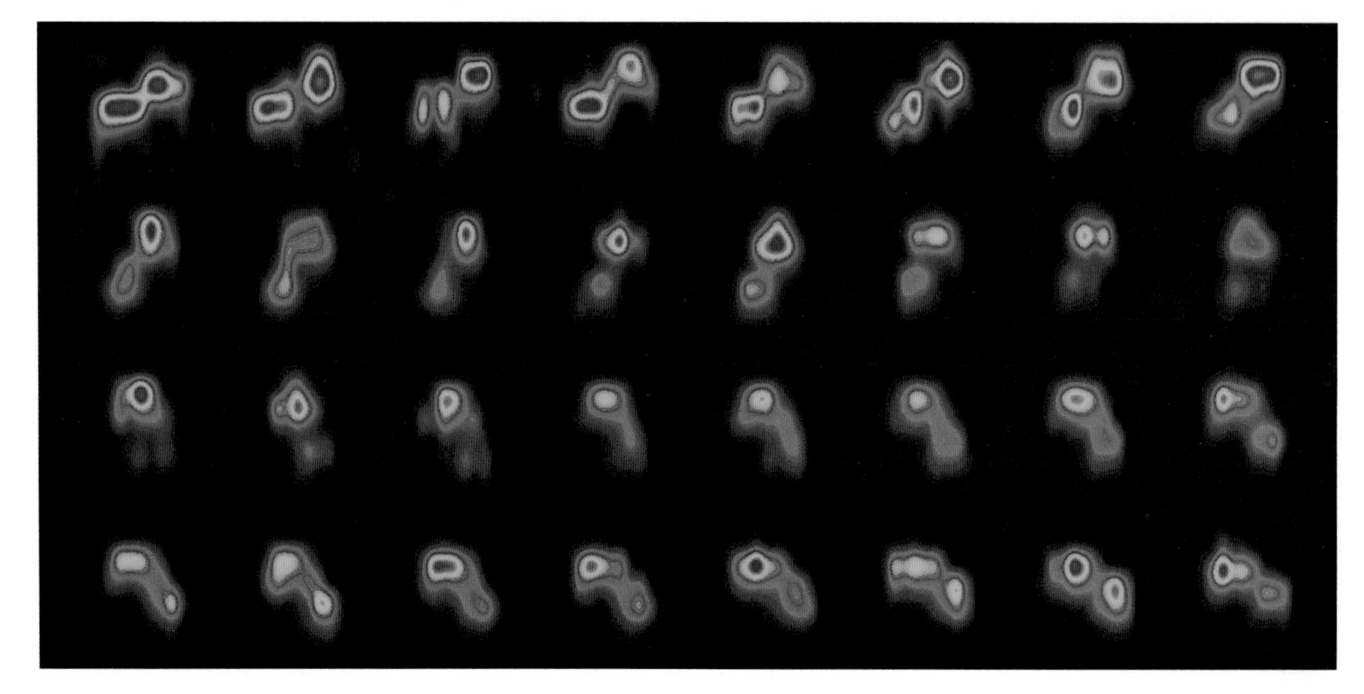

Radar images of the near-Earth asteroid Castalia reveal that this object consists of two lobes each about 500 m in diameter. Judging from these low-resolution images, Castalia may be almost dumbbell shaped.

While the minerals inferred for the asteroids are similar to those in many stony meteorites, exact identifications are difficult. The tools we have for studying the asteroids cannot assess a unique fingerprint. Our current knowledge is more nearly equivalent to the verbal description of a criminal suspect; we can categorize in terms analogous to height, weight, age, sex, and hair color, but we cannot specify the unique properties that identify an individual. We can conclude in general that the asteroids are the parent bodies of most meteorites, but there are few secure links between individual asteroids and individual meteorites.

One of the most convincing associations is between the large asteroid Vesta and a group of igneous meteorites called the eucrites. Vesta is unique among the larger asteroids in having a surface of basaltic lava, similar to the basalts found on the Earth or Moon. Detailed comparisons with the eucrites show that the lava flows on Vesta are apparently identical with those represented by these meteorites. It is therefore highly likely, although not proven, that Vesta is the parent body of this class of meteorites.

An asteroid family is defined as a group of objects with similar orbits, suggesting a common origin. Although not clustered together in space at present, the members of an asteroid family were all at the same place at some undetermined time in the past. Apparently the family members are fragments of broken asteroids, shattered in some ancient collision and still following similar orbital paths. According to some estimates, almost all of the asteroids smaller than about 200 km in diameter were probably so disrupted in earlier times, when the population of asteroids was larger. The families we see today may be remnants of the most recent of these interasteroid collisions.

Not all asteroids are confined to the belt between Mars and Jupiter. There are a number of dark, primitive asteroids at the distance of Jupiter, preceding and following the giant planet at an angle of about 60 degrees. These are called the Trojan asteroids; like those of asteroids in the belt, their orbital locations are stable. Asteroids also leak from the belt as a result of collisions and subsequent gravitational perturbations by the planets, especially Jupiter. Those that wander into the inner solar system are called near-Earth asteroids, or NEAs. Their orbits are not stable, and most of the NEAs will meet a catastrophic fate when they collide with one of the terrestrial planets. We will return to the NEAs in the Epilogue.

A Close-Up Look

On Halloween 1991, the Jupiter-bound Galileo spacecraft made the first close flyby of an asteroid – a small object named Gaspra. The spacecraft data for Gaspra provide our only real information on the appearance of these celestial fragments, although we have also learned about asteroids from earlier spacecraft measurements of Phobos and Deimos, the small satellites of Mars. Because these two small moons are probably captured asteroids, acquired by Mars shortly after its formation, they should be able to tell us something of their asteroid cousins.

Gaspra is a small stony asteroid from the inner part of the main belt, with an average distance from the Sun of 2.2 AU and an orbital period of 3.3 years. It is elongated, about 17 km long and 11 km wide, and is one of many members of the Hungaria asteroid family.

In its long and indirect flight path to Jupiter, the Galileo spacecraft was able to fly past two asteroids at close range: Gaspra in October 1991 and Ida in August 1993.

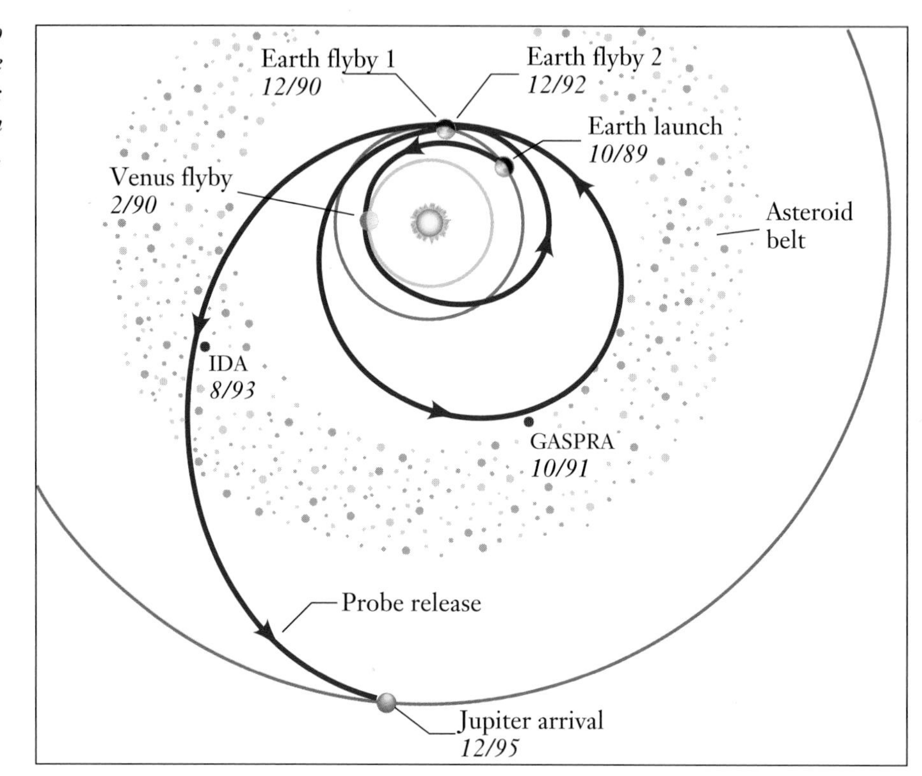

Gaspra was selected by chance for the first asteroid flyby; it was the easiest known asteroid for the Galileo spacecraft to reach on its predetermined path to Jupiter. Only a minuscule tweak of the spacecraft's flight path was required to bring it within 1600 km of the asteroid.

During its final hours of approach, the spacecraft revealed that Gaspra appears to consist of two lumps squashed together, the larger about 12 km across and the smaller about half this size. Its period of rotation is 7 hours. Almost surely the asteroid is a fragment, produced in some earlier collision that disrupted a larger preexisting asteroid. This may have been the catastrophic collision that broke apart the parent body and created the Hungaria asteroid family, or it might represent a subsequent collision involving one or more of the Hungaria family members. However, Gaspra's intriguing shape suggests to the Galileo project scientists that two separate fragments coalesced to form the strange object we see today.

From the accumulation of small impact craters on its surface, we can estimate how much time has transpired since the birth of Gaspra. Apparently this event (or events) took place about 200 million years ago, making Gaspra one of the younger members of the solar system. Based on the numbers of objects in the inner asteroid belt where Gaspra orbits, it is likely to experience another catastrophic collision sometime within the next billion years, at which time it will break up into many still smaller fragments.

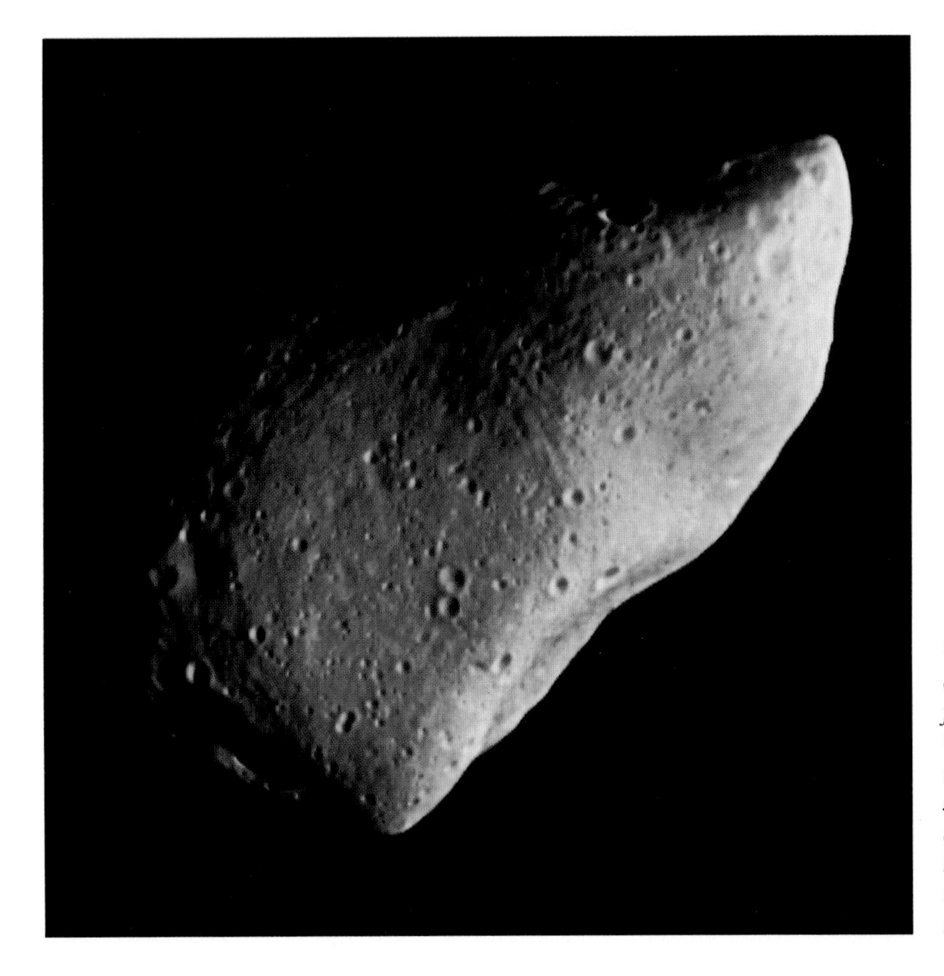

Gaspra is the first asteroid to be seen at close range. Photographed by Galileo from a distance of only 1600 km, Gaspra is revealed as an elongated, irregular object about 17 km in length, spinning with a period of 7 hours. The color distinctions shown here are real, but they have been highly exaggerated in this image; for most purposes, Gaspra has a neutral grey color.

The first three asteroidal bodies to be imaged from spacecraft are the main-belt asteroid Gaspra (top) and the two martian satellites Deimos (left) and Phobos (right). This image shows all three objects to the same scale and resolution, illustrating striking differences in their crater populations.

The captured martian satellites Phobos and Deimos are about the same size as Gaspra but of different composition. They are dark and primitive, like the carbonaceous meteorites. Phobos, the larger of the two (about 21 km diameter), is closest to Mars; its orbital period is less than 8 hours. Because Phobos revolves faster than Mars rotates, the satellite is seen (and was photographed by the Viking landers) to rise in the west and set in the east. Deimos, about half the size of Phobos, has an orbital period of 30 hours. Both satellites always keep the same side turned toward Mars, as our Moon does toward the Earth.

Late in the Viking mission to Mars, special efforts were made to obtain very high resolution photographs of the satellites. With careful navigation a flyby of Phobos at just 88 km distance was achieved in February 1977, followed in May by one of Deimos. This Deimos encounter, just 23 km above the surface, is the closest any planetary spacecraft has approached an object without actually hitting it. Close-up views of Phobos show that it is heavily cratered, like Gaspra, but in addition there are parallel grooves or valleys, typically several

Photographed at very close range by Viking, the surface of Phobos shows a remarkable set of grooves of unknown origin. Similar but less conspicuous grooves can also be seen on the highest-resolution Galileo image of Gaspra.

hundred meters across and several tens of meters deep. Apparently the grooves are fractures, related to the impact that formed the largest crater on Phobos. Indeed, this crater is so large relative to the size of the satellite that its formation must very nearly have broken Phobos apart. A few grooves are also seen on Gaspra, suggesting that groove formation may be a characteristic by-product of the extreme battering sustained by these small objects.

These first spacecraft encounters tell us a great deal about the physical properties and collisional history of individual small asteroids, but they do not provide much information on their chemistry. We can learn much more about composition by turning our attention to the comets.

Long-Tailed Comets

Comets have been known since antiquity. Approximately one comet visible to the naked eye appears each year, and a bright comet comes

Two views of Comet Halley are shown in this painting of its 1456 apparition. The picture indicates that the comet is causing great destruction, with the sky raining blood. There is considerable debate as to why comets had such a bad reputation in ancient times.

along an average of once per decade. Written records of these long-tailed wanderers go back to at least 1140 B.C. in the Middle East, and nearly as far in China. Yet for all of the myths associated with comets, people are often disappointed when they actually see one. Perhaps they expect a spectacular light show, something akin to a fireworks display. Instead they see a faint, nebulous patch of light not much bigger than the Moon and a great deal fainter. Nor is the comet distinguished by its motion, which is imperceptible unless you watch for many hours. To see a comet at all, you must escape the smog and light pollution of our cities, but even under a dark sky you may require binoculars for a good view.

When we look at a comet, we are observing primarily the extended atmosphere that is its trademark. The small, solid body from which this atmosphere is released is called the nucleus; this is the real heart of the comet, even though it is generally too faint to

Some Famous Comets

Name	Period	Significance
Great Comet of 1811	long	largest head observed (2 million km across)
Great Comet of 1843	long	brightest ever; visible in daylight
Daylight Comet of 1910	long	brightest of twentieth century
West (1976)	long	best recent comet; nucleus split
Swift-Tuttle	133 years	parent comet of Perseid meteors
Halley	76 years	studied in detail by spacecraft (1986)
Biela	6.7 years	broke up in 1846 and disappeared
Giacobini-Zinner	6.5 years	first spacecraft encounter (1985)
Encke	3.3 years	shortest known period

be seen. The atmosphere that surrounds the nucleus is called the head, and the long streamers sweeping away from the Sun are called the tail. No matter which way the comet moves, its tail points away from the Sun, driven outward by charged particles and solar radiation.

One astronomer whose name has been immortalized for his work on comets was Edmund Halley, a contemporary of Isaac Newton in late-seventeenth-century England. It was Halley who first realized that cometary orbits could be closed ellipses, with a given comet reappearing at regular intervals. He reached this conclusion from a study of the Great Comets of 1531, 1607, and 1682, all of which had similar orbits. Halley concluded that these were successive appearances of the same object, which he predicted would return in

Most comets are on highly elongated orbits. They can be seen only near perihelion when the Sun heats their surfaces to form a tenuous but extensive atmosphere and tail. As illustrated here, the tail generally points away from the Sun.

1758. Although Halley did not live to see his calculations verified, the comet returned just as he had expected and was quickly hailed as Halley's Comet, a name it retains today.

Comets have the most elongated orbits of any members of the solar system. Long-period comets fall toward the Sun from a great distance and return again to the depths of space. In addition, there are shorter-period comets that have been deflected into more circular orbits by interactions with Jupiter.

The most important part of a comet is its nucleus; the rest is insubstantial show. We know a nucleus must exist, since the transient gases of the comet's atmosphere have to come from somewhere. This nucleus is composed of a mixture of ice and dust, like a dirty snowball. As it approaches the Sun, the ice and other volatiles begin to evaporate, producing the atmosphere; because the gravitational pull of the nucleus is so weak, this gas expands to enormous size. The head and tail are what distinguish a comet and make it so easily visible, but the glare of the atmosphere obscures the tiny nucleus. Only recently has radar penetrated the atmospheres of several comets

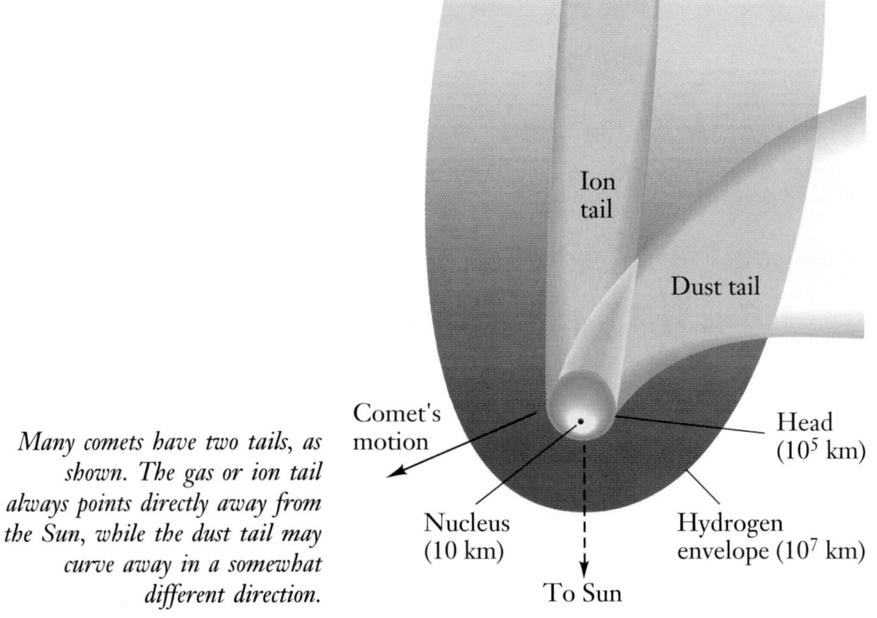

Many comets have two tails, as shown. The gas or ion tail always points directly away from the Sun, while the dust tail may curve away in a somewhat different direction.

to provide unambiguous detection of their nuclei, while the spacecraft that flew through Comet Halley in 1986 were able to photograph the nucleus of that comet at close range.

Comet Halley

During the week beginning March 6, 1986, three spacecraft flew into the heart of Comet Halley. First the Soviet spacecraft VEGA 1 and 2 approached to within about 8000 km and photographed a thick envelope of dust surrounding the nucleus, with several bright jets of gas and dust erupting from it. Using navigation data from the VEGA missions, the European Giotto spacecraft targeted a flyby only 500 km from the nucleus for its March 14 encounter. Its cameras pierced the dust fog and captured images of the nucleus itself, an irregular dark mass about 16 by 8 km in size, just a little smaller than asteroid Gaspra. The reflectivity of this nucleus was only about 3 percent, similar to the darker asteroids and indicative of primitive,

These views of Comet Halley were obtained from Tucson on January 20, 1986; they are two representations of the same data. On the left is the image in plain black-and-white, approximately as it appeared to the eye. The second image encodes the same data in false color to bring out fine details. False color or exaggerated color images are used increasingly in astronomy; however, in this book we have avoided false color and generally tried to use the most accurate and realistic photographs available.

carbonaceous material; the nucleus is a very dirty snowball indeed. Presumably the loss of ice and other volatiles from the upper layers of the nucleus allows a crust of dark dust to accumulate over the entire surface. A similar effect darkens the toe of a melting glacier on Earth by concentrating the dirt as the ice evaporates.

The spacecraft instruments measured water, carbon monoxide, carbon dioxide, and ammonia in the head of Comet Halley. They also discovered that the nucleus contains substantial quantities of dark carbonaceous and silicate dust. For this comet at least, carbon and hydrocarbon dust predominates over silicate dust. As the ices evaporate they release the dust particles, which stream into space, carried along by the flow of gas.

Perhaps the most spectacular features photographed on the nucleus were the bright jets of gas and dust erupting from the surface. Evidently evaporation does not take place from the entire nucleus but is focused into geyserlike eruptions. We can estimate the typical amount of gas and dust lost during one trip through the inner solar system as 10 to 100 million tons, corresponding to about 0.1 percent of the total cometary mass. Clearly, at this rate the comet will exhaust its store of ices after roughly a thousand passes through the inner solar system. The rocket effect of these gas jets also

This close-up image of the nucleus of Comet Halley was obtained in March 1986 by the European Giotto spacecraft. Prominent are several bright jets of gas and dust escaping from the dark, solid nucleus. The size and shape of the Halley nucleus are similar to those of asteroid Gaspra (page 195).

influences the orbits of comets in unpredictable ways through the action of so-called nongravitational forces.

While the dirty-snowball model provides a satisfactory understanding of many cometary phenomena, it also leaves important questions unanswered concerning details of the activity. What would it really be like to ride along on a comet as it approaches the Sun? The Halley probes seemed to indicate a highly dynamic environment as eruptions burst from different spots on the nucleus. Maybe ices other than water dominate the activity on some comets. Mysteriously, Halley brightened suddenly in 1991, when it had already retreated into the cold; we do not know what caused this event.

The Comet-Meteor Connection

Comets release quantities of dust or larger solid material into the inner solar system. A tiny fraction of this debris strikes the Earth, burning up in the atmosphere to produce the meteors.

Meteors — so-called shooting stars — can be seen on any clear night, usually at a rate of several per hour. The typical meteor is no larger than a pea, but this is sufficient to generate a much larger cloud of glowing gas high in our planet's atmosphere, visible from great distances. As many as 25 million meteors bright enough to be seen, amounting to hundreds of tons of cosmic material, strike the Earth's atmosphere every day.

Many meteors are produced by bits of material in random orbits. Sometimes, however, the Earth encounters a stream of particles moving together along similar orbits around the Sun, and we see a meteor shower. In a shower, all of the meteors appear to diverge from a single region of the sky. It is the shower meteors that are most directly linked to the comets.

To illustrate the connection between cometary dust and meteors more specifically, we can look to one of the famous comets of the last century: Comet Biela, discovered in 1826. In 1846 this comet split in two, and upon its next return in 1852 both components were again present, separated by 2 million kilometers. Neither part of Comet Biela was ever seen again; between 1852 and 1866, it simply ceased to exist. Nevertheless, astronomers watched with interest when

Major Annual Meteor Showers

Name	Date	Associated Comet
Quadrantid	January 2	unknown
Lyrid	April 21	Thatcher
Eta Aquarid	May 4	Halley
Delta Aquarid	July 30	unknown
Perseid	August 11	Swift-Tuttle
Draconid	October 9	Giacobini-Zinner
Orionid	October 20	Halley
Taurid	October 31	Encke
Leonid	November 16	Tempel-Tuttle
Geminid	December 13	Phaethon

Comets generate trails of dust ejected from the nucleus. The upper drawing illustrates the situation when this dust is concentrated in a part of the orbit. If the Earth passes through such a concentration we see an intense meteor shower; if we cross the orbit at some other location, few meteors will be observed.

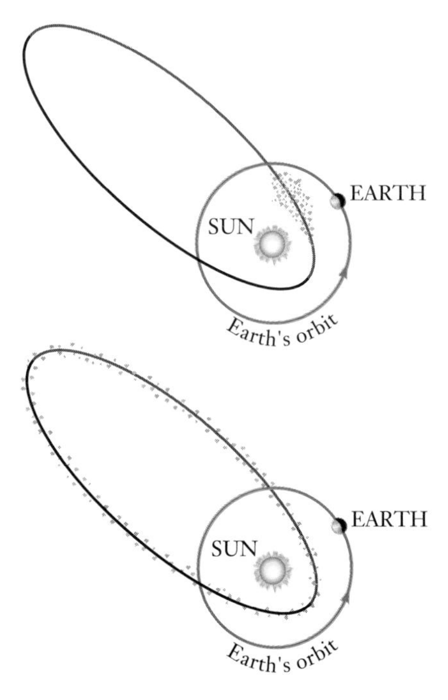

the Earth passed through the orbit of Comet Biela in 1872, and they were not disappointed. Instead of the comet they saw a wonderful meteor shower, with thousands of meteors visible from any spot on Earth during the night the Earth crossed the comet's orbit. The comet had transformed itself into a stream of dust.

The most dependable, although not the most spectacular, meteor shower can be seen each year between August 9 and 13. Called the Perseid shower, this stream of debris is associated with a bright comet with a 133-year period named Swift-Tuttle. Since the particles are fairly evenly distributed along the comet's orbit, we encounter about the same number each year. In contrast, the really spectacular but unpredictable showers occur when the Earth encounters a bunching of particles in the path of a meteor stream.

If most *meteors* are really cometary dust, we naturally ask if *meteorites* might not also be from comets. Meteorites are not associated with meteor showers, however, and even on the rare occasions when the sky is filled with "falling stars" no actual meteorites fall to Earth. The shower meteors are not simply small

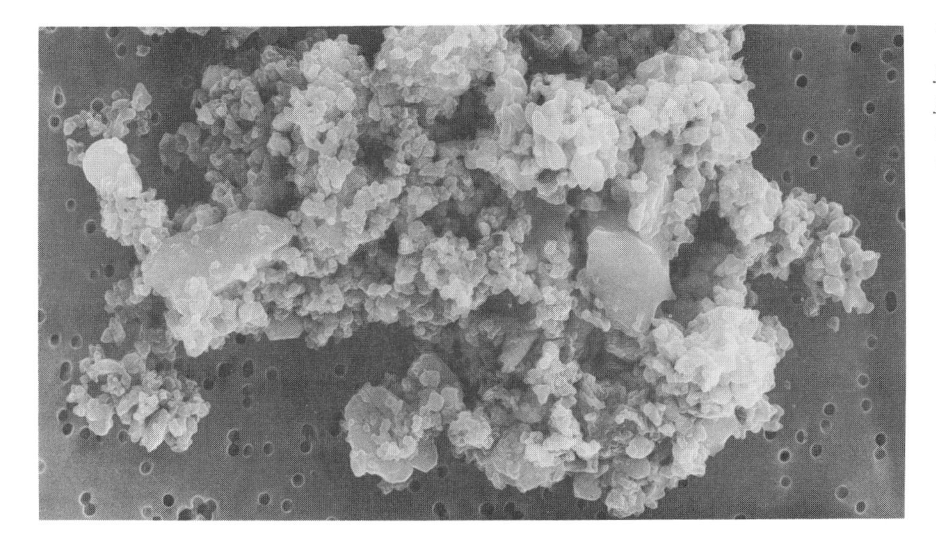

Tiny bits of comet dust can be collected from the Earth's upper atmosphere. This fragment, only a few micrometers across, is fluffy in structure and rich in organic (carbon-based) chemicals.

meteorites that do not reach the surface. They are fundamentally different material.

We have made some efforts to gather this comet dust, which is probably the most primitive material available for our direct study. Because they are so small, some of the dust grains that strike the Earth's atmosphere remain suspended in the stratosphere and can be collected from high-flying research aircraft and analyzed in the laboratory. The individual grains are fluffy and rich in carbon; there is, in addition, some chemical evidence for a component of interstellar material that predates the formation of the solar system. Important insights probably await the development of advanced technologies to collect comet dust outside the atmosphere, with minimum heating or physical damage. One suggestion has been to fly dust-collection devices past individual comets as an inexpensive way to return samples of cometary material for study on Earth.

Origin of Comets

The comets that we see are temporary residents of the inner solar system. After a few thousand orbits – less than 10,000 years, for many of them – their volatile ices will be exhausted and they will no longer be comets, although the dead nucleus may survive as a near-Earth asteroid (NEA). Comets are also dynamically unstable, however. Like all objects with planet-crossing orbits, they run the risk of either impact with a planet or gravitational expulsion from the solar system. Their average residency in the solar system is no more than a few hundred million years.

The first satisfactory theory for the origin of the comets was proposed by the Dutch astronomer Jan Oort in 1950. Oort noted that in all cases where the orbits of new comets had been carefully determined, the orbits indicated an aphelion at a distance of about 50,000 AU — a thousand times more distant than Pluto. Very few comets seemed to come from greater distances, and none showed evidence of originating outside the solar system in interstellar space. He therefore suggested the existence of a comet cloud associated with the Sun but very far beyond the known planets. This is now called the Oort comet cloud, and it is estimated to contain about a trillion comets; the total mass in the cloud may be as great as the mass of all the planets combined.

The comets of the Oort cloud may have formed in their present locations, but more likely they originated in the region of space now occupied by the outer planets. We expect large quantities of icy objects to have formed there. Many of these probably were incorporated into the giant planets, but still more may have been gravitationally ejected to much greater distances. This is the current theory, but no one knows for sure.

At any given time, virtually all of the comets in the Oort cloud are too far away for detection, and their ices will be preserved indefinitely in the cold of space. Even if the orbits of these comets bring them in as close as Neptune, they remain frozen and invisible to us. However, occasionally some of these comets can be perturbed by the slight gravitational tugs of nearby stars to bring them into the inner solar system. Only then will they be recognized as comets.

Once a comet is diverted into the inner solar system, its life expectancy is limited. There is some chance that it will not survive even its first plunge toward the Sun; or it may pass near enough to some planet — usually Jupiter — to suffer a change in orbit that either ejects it from the solar system or leads to its capture into a short-period orbit. If the latter happens, the comet has gone out of the frying pan and into the fire, for it still runs a significant risk of impact with one of the terrestrial planets. Meanwhile the heat of the Sun is consuming its icy substance at a rapid rate.

There is considerable debate among astronomers about the aging of comets. As a comet loses gas and dust each time it travels around the Sun, does it change or simply grow smaller? If it is of uniform, homogeneous composition, and if the dust is stripped away together with the evaporating ices, then we can expect the nucleus to shrink without otherwise changing much over time. Alternatively, if comets have a layered structure, the successive stripping away of the surface (typically about 1 m each orbit) will expose differing materials as the comet ages. To date there is no evidence for such lack of homogeneity, since the spectra of old comets are not different from those of young ones. A third possibility is that, while the nucleus is homogeneous, the non-volatile debris that builds up on the surface ultimately insulates the interior under a thick crust of dark material.

The Solar Nebula

With our description of the comets, we come to the end of our inventory of the solar system. These messengers from the deep freeze of the Oort cloud, together with the primitive asteroids between Mars and Jupiter, provide our strongest links to the time of formation of the planets. Let us now pull together the information we have to try to characterize the birth of the solar system.

Any theory of planetary origins should be consistent with the uniform age of the Sun, comets and asteroids, and the planets themselves, all of which formed together about 4.5 billion years ago. It should explain the observations that the planets all orbit in the same direction and in approximately the same plane and that their direction of motion is the same as the Sun's direction of rotation. Finally, it should incorporate the detailed information now available on the chemical composition of the various members of the solar system.

As we have noted throughout this book, all of our information on the planets can be understood if the Sun was accompanied at its birth by a disk of gas and dust that formed with it out of a collapsing interstellar cloud. This disk, called the solar nebula, was formed of material from the cold of interstellar space, but heated by self-compression as it collapsed into a spinning disk. Additional heat was

Star formation is taking place today in molecular clouds such as this, located at a distance of 7000 light years. Many young stellar objects are surrounded by disks of dust and gas that are analogous to the solar nebula that accompanied the Sun at its birth 4.5 billion years ago. Are these dust disks the nurseries for a new generation of planets?

supplied to the inner parts of the solar nebula from the protosun beginning to shine at its center. Throughout much of the nebula, temperatures were high enough for the interstellar dust to evaporate and mix with the gas. Then, as temperatures subsequently declined, solid material condensed from the cooling gas and accumulated to form the building blocks of the planets. These building blocks, typically a few kilometers in diameter (like many comets and asteroids today), are called planetesimals.

The planetesimals were closely packed in the rotating disk, where they interacted with each other and with the surrounding gas (primarily hydrogen and helium). Collisions fragmented some of these planetesimals, while others coalesced as a result of their mutual gravitational attraction. At distances greater than about 4 AU from the center, a few very large objects formed, with masses about 10 times that of the Earth today. These massive solid cores were able to attract and hold the surrounding nebular gas; they became the four giant planets. Each of these giant planets in turn developed its own small disk of surrounding material, and these gave rise to the major satellites of the outer solar system. At some point after the formation of the giant planets, the residual gas was stripped from the system and the planets were free to continue on their way; the solar nebula had given birth to the planetary system.

Within about 4 AU of the Sun, temperatures never dropped low enough for water ice and other common volatiles to condense, so the planetesimals were composed primarily of silicate rock and metal. Average speeds were higher here also, impeding the formation of objects larger than Earth or Venus. Eventually the four present terrestrial planets emerged, but only after a period of catastrophic collisions. One of these blasted material from Earth to form our Moon; another removed most of the silicate mantle from Mercury; while a third probably reversed the direction of spin on Venus. A continuing rain of icy planetesimals from the outer solar system provided a veneer of volatiles and organic material to the terrestrial planets – material that would ultimately provide an environment for the origin of life on Earth and, perhaps, on Mars as well. Most of the planetesimals either hit the planets or were ejected into the Oort cloud or beyond. A few small rocky objects found stable orbits between Mars and Jupiter, becoming the asteroids.

This description is, of course, only meaningful if it can be supported by detailed, quantitative models of the evolution of the solar nebula into the planetary system we see today. Such models exist, but there are many disputes concerning the details of the process. One way to check these ideas is to search for similar processes taking place today. Across our Galaxy, interstellar clouds are collapsing and new stars are being born. While we have not yet detected other planetary systems like our own, we have seen examples of the processes associated with our birth. Infrared observations have identified many young stellar objects surrounded by disks of infalling dust and gas, just as we imagine the solar system might have looked 4.5 billion years ago. The construction of new telescopes and dramatic improvements in spatial resolution should permit astronomers in the next few years to probe more deeply into these protoplanetary disks. Perhaps such studies – as much as continuing research on our own solar system – will provide the critical information needed to confirm or replace the current paradigm for the origin of the planets, Earth, and life itself.

EPILOGUE

COSMIC IMPACTS

A Planetary Perspective

...

Cosmic debris strikes the Earth every day, as depicted in this painting by Albrecht Dürer of an exploding fireball or bolide. Small, frequent impacts are harmless, but the very rare, large projectiles pose a significant threat to human civilization.

Earth orbits the Sun within a swarm of near-Earth asteroids (NEAs) and comets, as we saw in the previous chapter. Repeatedly in this book we have noted that impacts with such objects produce craters on the terrestrial planets and the satellites of the outer planets. Even today, this bombardment continues. How have these crater-forming impacts with comets and asteroids influenced the history of our world? Has our planet evolved in relative isolation, or was it shaped by its cosmic environment?

Until recently the idea that impacts or other global catastrophes influence history was the prerogative of pseudoscience. Under the umbrella of so-called creation science, Biblical literalists posit a worldwide deluge just a few thousand years ago that destroyed virtually all life – a myth that can be traced to roots in early Mesopotamia. In Nazi Germany, a cult of ice-catastrophists flourished, while more recently Immanuel Velikovsky attracted many followers in the United States with his book *Worlds in Collision.* None of these ideas has been based on scientific evidence, however, either terrestrial or cosmic, and most such catastrophism is founded in prescientific concepts of a young Earth and a cosmos that revolves, at least figuratively, around a central human presence.

Several recent discoveries have resurrected catastrophism as a scientific idea, at least in terms of the role of impacts on the Earth. As we have seen, astronomers are finding many comets and asteroids whose orbits cross that of the Earth, and calculations show that such objects must occasionally strike our planet. The exploration of the solar system by spacecraft has taught us that impact cratering is a ubiquitous component of planetary evolution. Who can look at the cratered face of the Moon and not recognize that the Earth has been subject to a similar flux of projectiles?

Does the formation on Earth of a large crater every few million years really make any difference? Not to the planet as a whole. Such impacts are minuscule on a global scale; they do not affect our planet's orbit or rotation or change the tilt of its axis. We would not care about such events if ours were a dead world like the Moon. What makes such impacts interesting—and potentially catastrophic—is their ability to influence the evolution of life.

In this Epilogue I discuss recent evidence for the importance of impacts in the history of life—future history as well as past. These ideas help us to place our planet in a cosmic context and understand its links to other members of the solar system. They also represent a particular interest of mine. Impacts raise important public policy questions, since we are in a position to protect Earth from most future impacts if we choose to do so. I find the prospects for influencing the future evolution of life both exciting and disturbing. Most other planetary scientists would not give such prominence to impact risks in a book on the planets, but I think these concepts are a worthy way to conclude this volume.

Impacts and Extinctions

Sixty-five million years ago the Earth was very similar to the planet we know today. Indeed, this period is short on the cosmic clock, little more than 1 percent of the age of the solar system. If you looked down on our planet then, you could recognize most of the major continents and oceans, although Antarctica, Australia, and the Indian subcontinent were not quite in their present locations. The atmosphere and climate resembled today's conditions; a little warmer perhaps, but nothing dramatic. Only if you inspected the life on our planet would you see a major difference, for the dominant large animals on land, in the sea, and in the air were dinosaurs, one of the most successful groups of animals to emerge on our planet. Dinosaurs had flourished for more than 100 million years and seemed destined to continue their dominance forever. But their world was about to end, as the result of a cosmic collision.

An object about 15 km in diameter, with a mass of 10 trillion (10^{13}) tons, was perturbed by the gravitation of other planets into a

The sudden extinction of the dinosaurs has intrigued scientists and laypersons alike for more than a century. Now we are confident that the primary cause of this extinction was one or more cosmic impacts. Could the same thing happen to us?

collision course with the Earth. This object may have been a rocky asteroid or a comet; there is some evidence that it was just one member of a stream of cometary fragments formed when a much larger object broke up as it passed close to the Sun. Possibly there were a number of close passes by Earth before the final impact. Eventually, however, there came the day when the Earth lay in the inevitable path of the projectile.

Suddenly, almost without warning, the object plunged into our atmosphere above the Yucatan peninsula in what is today southern Mexico. At a speed of at least 25 km/s, it roared through the atmosphere in less than 3 seconds and smashed into the shallow seas of the North American continental shelf. Only a single second was required for it to bury itself in the crust and come to a halt, releasing a billion megatons of energy—nearly a million times the total explosive force of all the nuclear weapons in the arsenals of the world. This awesome energy shattered the rock of the crust for hundreds of kilometers and burst upward as an incandescent fireball of vaporized rock, blasting a crater 180 km in diameter and more than 20 km deep. The planet shuddered under the seismic hammer blow. A tsunami hundreds of meters high swept across the

Caribbean, devastating Hispaniola, Cuba, Florida, and the coast of Mexico. Heat from the explosion ignited the forests of Mexico, Venezuela, and Colombia to generate a firestorm of unprecedented magnitude. Within 5000 km of the impact lay only death and devastation.

Today we have located the remnant of this crater – called Chicxulub Crater after the Yucatan fishing village near its center – buried under a thick layer of limestone. We can trace the path of the tsunami in the Caribbean and find the deposits of material ejected from the crater. The explosion itself, however, did not lead to the vanishing of the dinosaurs or the other myriad species that mark this as one of the greatest mass extinctions of life in the fossil record. The reason we are interested in the Chicxulub impact is its ability to affect climate on a global scale.

The Chicxulub impact excavated more than 1000 trillion (10^{15}) tons of material, a partially vaporized mix of the shattered projectile and the crustal rock of the Earth at the point of impact. Most of this ejecta fell back near the crater, but about 1 percent, in the form of fine dust, was injected into the stratosphere where it remained suspended for months while winds carried it around the planet. The result was an opaque, black cloud that shrouded the Earth's surface from the light and heat of the Sun. For months the planet was cast into an endless winter night as land temperatures fell below freezing and photosynthesis ceased even in the oceans. The cold and darkness killed almost indiscriminately on a global scale. The fossil record shows that most of the species on Earth perished utterly, apparently including the dinosaurs. When the dust settled and the climate returned to normal, perhaps after a few years, much of the land and sea was barren of life – a situation ideal for the proliferation and rapid evolution of the surviving species.

As mammals we owe our present dominant status to the fact that our ancestors were among the survivors of the Chicxulub impact. Generally, the animals that avoided extinction were of small size and widely distributed population; burrowers on land and bottom-feeders in the seas were favored. The rodentlike mammals of 65 million years ago fit this description, possessing the attributes to survive the once-in-a-hundred-million-years cosmic catastrophe. Apparently the dinosaurs were not so fortunate.

This description of the environmental consequences of the Chicxulub impact is based on direct evidence. The global pall of dust that led to the mass extinction is still present in the strata of sediment laid down 65 million years ago, at the boundary between the Cretaceous and Tertiary eras of geologic history. Everywhere on the globe where this part of the geologic record survives, we see a distinct layer a few centimeters thick made up of fine dust and the characteristic mineral grains that are formed only in a great explosion. The extraterrestrial cause of the explosion is witnessed by the presence in this layer of excess quantities of rare elements such as iridium that are virtually absent in the Earth's crust (because they are in the core of our differentiated planet) but more common in primitive asteroids and comets. Even after dilution by mixing with material excavated from the crater, this extraterrestrial signature remains to be read in the sedimentary rocks laid down 65 million years ago.

Direct evidence of the impact 65 million years ago that destroyed the dinosaurs is found in the boundary layer between the sediments of the Cretaceous and Tertiary periods. This boundary clay includes anomalous quantities of the element iridium, which can have been derived only from an extraterrestrial source, as well as microspherules of impact-generated glass.

The iridium anomaly in the Chicxulub sediment layer was discovered nearly a decade before the impact crater itself was identified. A team of Berkeley scientists led by the Nobel physicist Luis Alvarez and his geologist son, Walter, found the iridium anomaly and suggested in 1980 what now seems to be the correct explanation for both the origin of this layer and its association with the mass extinction that occurred at the end of the Cretaceous period.

The immediate reaction of most paleontologists to the Alvarez hypothesis was negative. Such a scenario challenged their basic assumption that evolution proceeded slowly by small increments in response to gradual changes in the environment. Many remain hostile a decade later. However, the accumulation of evidence has convinced most scientists that the story we have recounted of the Chicxulub impact and its consequences for life is essentially correct. Because its effects led to a short-term climate change, this impact dramatically changed the course of evolution.

Impacts of comets or asteroids 10 km or larger in diameter are expected every few tens of millions of years. Those as large as the Chicxulub event come at intervals of about 100 million years on average. Perhaps by coincidence, the typical interval between mass extinctions indicated in the fossil record is about 30 million years. In the 250 million years since the end of the Permian, there are six well-documented mass extinctions. Is it possible that all of these events are impact-related? In no other case is the impact hypothesis as well supported as for the end-Cretaceous extinction, but some scientists argue that such impacts provide the most likely explanation for all of these discontinuities in the evolution of life.

If impacts are the dominant cause of extinction, and especially of the dramatic mass extinctions that represent the turning-points of evolutionary history, then our planet has been profoundly influenced by its cosmic environment. Indeed, the ability to survive an impact catastrophe becomes a prime requirement for long-term evolutionary success. We are here today only because our ancestors were able to withstand a series of such sudden perturbations in climate. In this sense at least, history is driven not by internal forces, but by the stars.

Cosmic Impact Hazard

If it happened before, it can happen again. Comets still descend unannounced from the Oort cloud, and collisions in the asteroid belt continue to eject fragments into Earth-crossing orbits. Small impacts (producing bright meteors and occasional meteorites) take place daily. Although the probability of a large impact is low, such an event could happen at any time. Like the dinosaurs, we might be caught by surprise, since even today we do not keep a sharp watch on the space surrounding the Earth.

In 1908 an asteroidal fragment with a mass of about 100,000 tons plunged, early in the morning of June 30, into the atmosphere above the Stony Tunguska River of Siberia, penetrating to within less than 10 km of the surface before it disintegrated from the stress of its rapid deceleration. The result was an "airburst" explosion with 10 to 20 megatons of energy, which flattened the forest to a radius of about 30 km. Because the region was sparsely inhabited, there were no casualties, but such an event in a populated area could destroy a large city and kill millions of people.

There is no historical record of any person being killed by a meteorite. Although stones fall from the sky, they generally do not

The largest historic impact took place on June 30, 1908, in the remote Tunguska region of Siberia. The incoming rocky meteorite exploded at an altitude of about 10 km with the force of 10 to 20 megatons of TNT. As shown in this photograph, the forest had not recovered from this blast several decades later.

strike with crater-forming energies, and they are so rare that they pose a negligible hazard. No one is on record as dying from a large impact either, but on a statistical basis such impacts pose a hazard that can be calculated and evaluated just like the risks of other natural phenomena like earthquakes and hurricanes.

Small impacts are much more frequent than large ones, reflecting the size distributions of the cometary and asteroidal projectiles; this is why there are many more small craters on the Moon than large ones. The largest lunar craters correspond to impacts of approximately the magnitude of the Chicxulub event on Earth, and the largest currently known Earth-crossing asteroids and comets are about 15 km in diameter. Impacts by objects this size are rare but devastating. Smaller impacts are more common but individually less destructive. A careful analysis is required to determine which is more risky: the rare big hits or the more frequent small ones.

Earth's atmosphere protects us from most impacts by objects less than about 100,000 tons, corresponding to a diameter of tens of meters. The stresses of passage through the atmosphere lead to fragmentation and sudden disruption of smaller objects before they can strike the surface. If the projectile is cometary and hence composed in part of ice, it disintegrates at altitudes above 50 km, and we record it simply as a very bright meteor, or bolide. Rocky objects can penetrate deeper, and the strongest of these, like the Tunguska projectile, will cause damage; however, in this size range only iron objects, which are thought to constitute less than 2 percent of Earth-crossing material, will reach the surface and produce craters. On average, we can expect a Tunguska-class atmospheric explosion somewhere on Earth once every few centuries, while a crater like Meteor Crater in Arizona is formed on land perhaps every few tens of millennia.

Despite their relative frequency, most Tunguska-class events will take place over oceans or other unpopulated regions and do little harm. Even if the explosion occurred over a populated region, the blast effects would be restricted to the local area around the target. There are no global climate perturbations and no evolutionary consequences. If you happened to be near the target, the situation could be injurious or even fatal, but the remaining 99.999 percent of the human population would not be at risk. The annual risk to any individual of dying from a Tunguska-class impact is less than 1 in 10 million.

Meteor Crater in northern Arizona is one of the best preserved impact structures on our planet. It was formed about 50,000 years ago from the impact of an iron meteorite with an energy release of about 15 megatons. The crater has a diameter of a little more than 1 km.

About 150 ancient impact structures have been identified on the Earth. One of the largest is Manicouagan Lake in Quebec, Canada. The original crater, with a diameter estimated at 65 km, was formed about 210 million years ago. This orbital photograph was taken by the Landsat remote-sensing satellite.

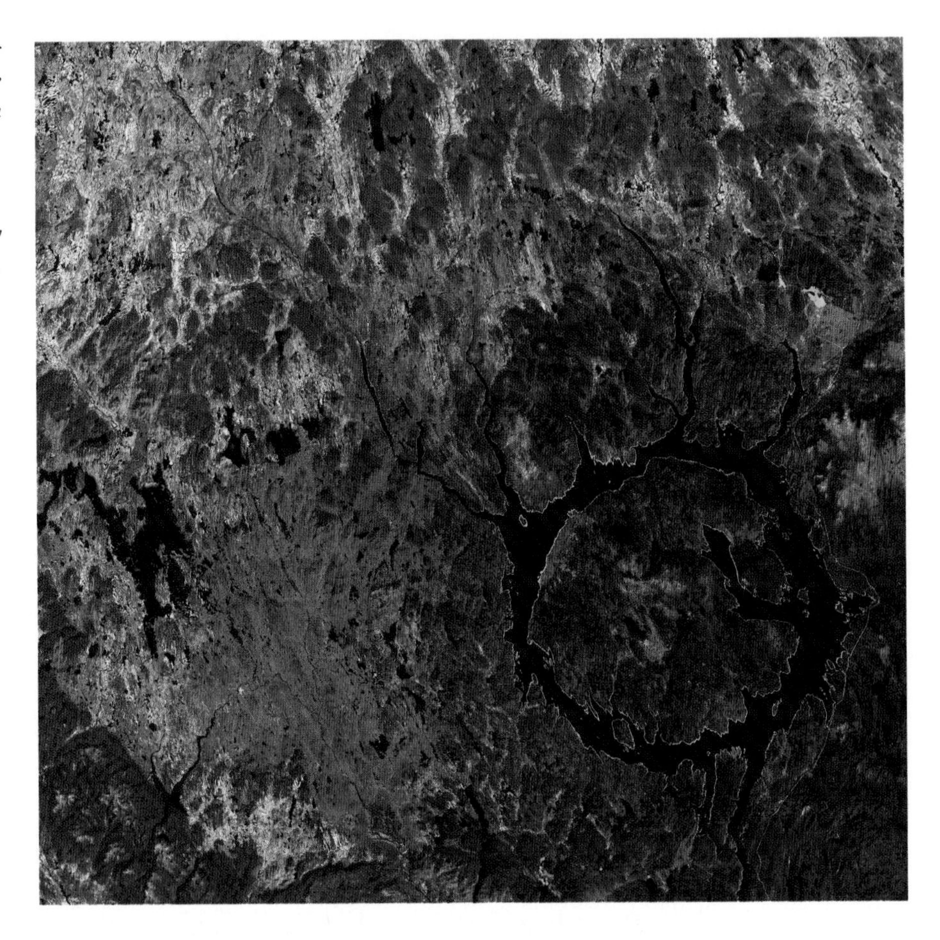

We have seen that the impact of a 15-km object will so severely affect the climate that it produces a mass extinction. A smaller impact can have similar effects on a smaller scale. Suppose that the stratospheric dust were only great enough to cool the surface by a few degrees for a few months. The result is not a mass extinction but a global crop failure. Humans depend on continuing high yields from grain crops, and there is little margin for error. No species would become extinct from such an impact, but a billion or more people could die of starvation. An object as small as 1 to 2 km in diameter could probably produce such an "impact winter." Objects of this size strike once every few hundred thousand years on average. If this analysis is correct, such objects pose the greatest risk.

Although there is no record of any human death from a meteorite fall, there are several documented cases of cars being struck by small meteorites. This meteorite fell in rural Illinois in 1938; it penetrated the garage roof, the roof of the automobile, and its seat cushion before finally coming to rest.

Suppose an impact that produces global crop failure takes place once every 250,000 years, and that if such an event happens during your lifetime you have a 1 in 4 chance of dying from impact-induced starvation or disease. Then your annual chance of death from this cause is 1 in 1 million. As the human lifetime approaches a century, this means that the lifetime risk of death from impact is about 1 in 10,000; that is, the chances are about 1 in 10,000 that any individual will die from the results of an impact rather than some other cause, natural or accidental. This risk can be compared with other hazards that we face.

Some forms of accidental death are much more likely than impact mortality, of course. In the United States at the end of the twentieth century, each individual's chance is about 1 in 100 of dying from an auto accident, about 1 in 200 of being shot to death. Compared to the risk of cars or guns, comets and asteroids are quite benign. Impact risks seem more substantial, however, when compared with other natural hazards. If you are an average American, your risk of dying from the consequences of an impact is greater than the combined risk of death from tornadoes, hurricanes, earthquakes, forest fires, and volcanic eruptions. Most people think it appropriate to take some precautions, either personal or governmental, against these hazards. By the same arguments, we should also worry about impacts — or at least think about how we might protect ourselves against such global catastrophes.

Protecting Earth

No one has figured out how to hush a hurricane or tame a tornado, but we do have some ideas on how to avoid asteroid and comet impacts. Prevention is the key: we must discover the projectiles while they are still far from the Earth and deflect them so they miss our planet.

As we have noted above, the greatest risk is associated with impacts by objects 1 km or more in diameter. Most of these are near-Earth asteroids (NEAs) with orbits that intersect that of our planet; as of 1992, astronomers have discovered 110 such asteroids with diameters of at least 1 km. Fortunately none of these objects poses any danger of impact for the foreseeable future, although one known comet, called Swift-Tuttle, has an orbit that brings it so close to Earth that there is a small but significant chance of impact when it next returns to the inner solar system in the summer of 2126. We can estimate that there are a total of about 2000 NEAs and short-period comets with Earth-crossing orbits, and it is impossible to make predictions about the orbits of the 95 percent that have not yet been discovered.

In a program to protect against asteroid impacts, the first step would be a survey to discover the two-thousand-odd objects greater than 1 km in diameter that have the potential to strike the Earth. Current searches, which use small telescopes and occupy only a handful of astronomers, are turning up about 10 per year. At this rate, it will require two centuries to obtain an inventory. Substantial acceleration of the discovery rate requires bigger telescopes, sophisticated automatic data processing, and more people — in other words, money. To compress a relatively complete survey from two centuries to two decades of work would cost about $100 million — a sum that could be shared by a number of technologically advanced nations, since everyone on the planet is equally at risk.

Most likely by far is that such a survey would find no NEAs that threaten Earth on a time scale of centuries. But suppose we did locate a potentially deadly object. In all probability the discovery would be made many decades before impact. Using optical and radar tracking, we could refine the orbit of the asteroid; small robotic spacecraft could fly out to measure its size, shape, rotation, and composition. Armed with this information, we would develop a strategy to protect our planet from disaster — probably a series of missions to alter the orbit of the asteroid. If we picked the appropriate part of its orbit, we could move it into a nonimpact trajectory with just a gentle nudge — typically a change in speed of only a few centimeters per second, or about 0.0001 percent of its orbital velocity. One way to accomplish this would be by detonating a modest neutron bomb near the NEA, heating the surface and boiling off enough rock vapor to induce a small reaction force on the asteroid, similar to the nongravitational forces on comets mentioned in the previous chapter. Given a lead time of several decades, the process of modifying the orbit could be carried out in small steps under international scrutiny.

Comets are more of a problem. Descending into the inner solar system from the Oort cloud, a new comet cannot be discovered decades before it threatens the Earth. Indeed, even with a comprehensive sky survey, the warning time for a comet could be as little as 12 months. The interception and deflection maneuver must be performed close to the Earth, requiring more powerful rockets and much larger nuclear explosives than those needed to deflect an asteroid. Further, the nongravitational forces on a comet introduce so much orbital uncertainty that we might not be confident whether it posed a real threat of impact at the time an interception decision needed to be made. This is the problem with Comet Swift-Tuttle, mentioned above; even the best current observations leave an uncertainty of several weeks in its 2126 return — a range that includes the possibility of Earth impact. Given these difficulties, we are fortunate that cometary impacts are much less frequent than those from asteroids.

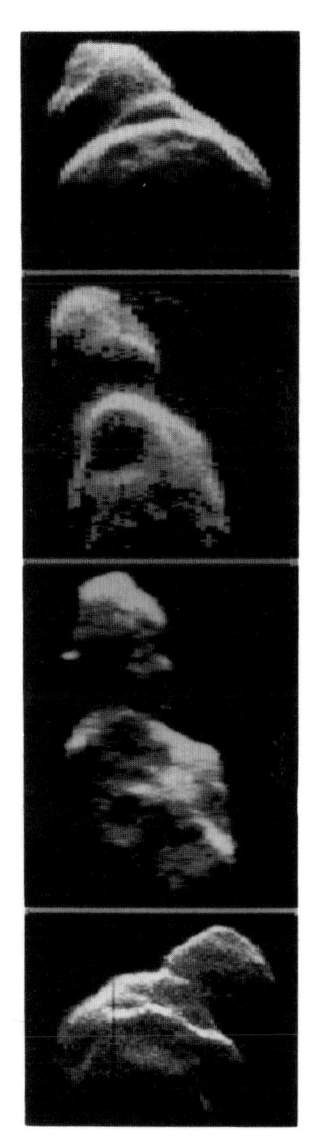

The first radar images to clearly resolve features on an asteroid were obtained over a 5-day interval in December 1992, when NEA Toutatis passed within 4 million km of the Earth. Toutatis consists of two objects, with diameters of about 4 and 3 km, apparently in contact with each other. This "double asteroid" rotates with a period of about 11 days.

The concept of catastrophic impacts is not new, as illustrated by this nineteenth-century French cartoon of the Earth being destroyed by a comet. Today, however, we have a much better understanding of the nature of the impact threat, and there are even prospects for averting future asteroidal and cometary impacts.

A Planetary Perspective

There is a wide range of opinion on the importance of the impact hazard. Mounting a telescopic search for NEAs is fairly inexpensive and noncontroversial. Discussions of interception, however, especially if we wish to respond on short notice, raise concerns about the new space arms that might be required. Some people think the risk of developing these weapons would be greater than the risk of the impacts themselves.

On a statistical basis, as we have seen, the risk to an individual of death from impacts is comparable to the risk of mortality from other natural hazards such as earthquakes and storms – on the order of 1 chance in 1 million in any given year. Some people worry about hurricanes and earthquakes, but most find it easy to dismiss the danger altogether when the odds are so low. Besides, no one in human history has died of an impact, while we read every year of deaths from other kinds of natural hazards.

One reason for concern is the truly catastrophic nature of large cosmic impacts. Unlike any of the other hazards we have considered, an impact could lead to more than a billion deaths worldwide and threaten the stability of civilization. Much more than loss of life is therefore at risk. This qualitative difference provides an added motivation to take action.

Facing the impact hazard requires us to develop a planetary perspective. Impacts by comets and asteroids are one way to disrupt the natural balance of the entire planet, and examination of these risks can sensitize us to other, human-caused threats to the delicate equilibrium of the Earth's biosphere. Two such threats are global warming, resulting from the excessive burning of fossil fuels, and depletion of stratospheric ozone, primarily from release of industrially produced CFCs.

The risks associated with global warming and ozone depletion are more difficult to calculate than those of catastrophic impacts. Neither is likely to produce the human deaths that would result from an impact winter and associated sudden loss of crops. We can expect the world to recover from an impact winter within a few years, however, while the effects of global warming and ozone depletion are much longer lived. Even if we stopped production of CFCs tomorrow, it would require more than a century for the atmosphere to cleanse itself. In the case of the release of carbon dioxide into the atmosphere, we do not understand the carbon budget of the Earth well enough to predict the consequences. Meanwhile human deforestation and desertification may already be leading to the extinction of half the species on our planet. Global warming and ozone depletion are real problems that confront us today.

Perhaps these concerns about the global environment, together with awareness of the potential threat of impacts, will generate a broader consciousness of the Earth as a planet. We now recognize that environmental balance is a delicate thing. In Venus and Mars we see possible endpoints in the evolution of terrestrial planets: one baking under a massive greenhouse, the other frozen in a terminal ice age. An important reason for studying the solar system is to gain a greater appreciation of the uniqueness of the Earth and an awareness of the possibilities for unexpected and undesirable change on our own planet.

FURTHER READINGS

The following recommended contemporary books on the solar system range from semi-popular to semi-technical.

GENERAL SOLAR SYSTEM

The New Solar System, Third Edition, edited by J. Kelley Beatty and Andrew Chaikin. Sky Publishing Corporation and Cambridge University Press, 1990.

The Cambridge Photographic Atlas of the Planets, by G.A. Briggs and F.W. Taylor. Cambridge University Press, 1982.

Exploring Space: Voyages in the Solar System and Beyond, by William E. Burrows. Random House, 1990.

Solar System, by Kenneth Frazier. Time-Life Books, 1985.

The Search for Life in the Universe, Second Edition, by Donald Goldsmith and Tobias Owen. Addison-Wesley Publishing Company, 1992.

Exploring the Planets, by W. Kenneth Hamblin and Eric H. Christiansen. Macmillan Publishing Company, 1990.

Moons and Planets, Third Edition, by William K. Hartmann. Wadsworth Publishing Company, 1993.

Bound to the Sun: The Story of Planets, Moons, and Comets, by Rudolf Kippenhahn (translated by Storm Dunlop). W.H. Freeman and Company, 1990.

Echoes of the Ancient Skies, by E.C. Krupp. Harper and Row, 1983.

The Planetary System, Second Edition, by Tobias Owen and David Morrison. Addison-Wesley Publishing Company, 1993.

Journey into Space: The First Three Decades of Space Exploration, by Bruce C. Murray. W.W. Norton and Company, 1989.

Cosmos, by Carl Sagan. MacDonald Futura Publishers, 1981.

EARTH, LIFE, IMPACTS

Cosmic Catastrophes, by Clark R. Chapman and David Morrison. Plenum Press, 1989.

Cosmic Impact, by John K. Davies. St. Martin's Press, 1986.

Meteorite Craters, by Kathleen Mark. University of Arizona Press, 1987.

Early Life, by Lynn Margulis. Jones and Bartlett Publishers, 1984.

The Nemesis Affair: A Story of the Death of Dinosaurs and the Ways of Science, by David M. Raup. W.W. Norton and Company, 1986.

Extinction: Bad Genes or Bad Luck?, by David M. Raup. W.W. Norton and Company, 1991.

The Genesis Strategy: Climate and Global Survival, by Stephen H. Schneider with Lynne E. Mesirow. Plenum Press, 1976.

The Coevolution of Climate and Life, by Stephen H. Schneider and Randi Londer. Sierra Club Books, 1984.

Landprints, by Walter S. Sullivan. Times Books, 1985.

TERRESTRIAL PLANETS

(Note: No books yet out reflect recent results on the geology of Venus.)

The Surface of Mars, by Michael H. Carr. Yale University Press, 1981.

The Geology of the Terrestrial Planets (NASA SP-469), by Michael H. Carr, R. Stephen Saunders, Robert G. Strom, and Don E. Wilhelms. National Aeronautics and Space Administration, 1984.

Planets of Rock and Ice: From Mercury to the Moons of Saturn, by Clark R. Chapman. Charles Scribner's Sons, 1982.

Apollo on the Moon, by Henry S.F. Cooper, Jr. The Dial Press, Inc., 1969.

The Search for Life on Mars: Evolution of an Idea, by Henry S.F. Cooper, Jr. Holt, Rinehart, and Winston, 1980.

Apollo Expeditions to the Moon (NASA SP-350), edited by Edgar M. Cortright. National Aeronautics and Space Administration, 1975.

To Utopia and Back: The Search for Life in the Solar System, by Norman H. Horowitz. W.H. Freeman and Company, 1986.

Apollo Over the Moon: A View from Orbit (NASA SP-362), by Harold Masursky, G.W. Colton, and Farouk El-Baz. National Aeronautics and Space Administration, 1978.

Earthlike Planets: Surfaces of Mercury, Venus, Earth, and Mars, by Bruce Murray, Michael C. Malin, and Ronald Greeley. W.H. Freeman and Company, 1981.

Mercury: The Elusive Planet, by Robert G. Strom. Smithsonian Institution Press, 1987.

Lunar Science: A Post-Apollo View, by Stuart Ross Taylor. Pergamon Press, Inc., 1975.

Mars at Last!, by Mark Washburn. G.P. Putnam's Sons, 1977.

OUTER PLANETS

Far Encounter: The Neptune System, by Eric Burgess. Columbia University Press, 1991.

Flyby: The Interplanetary Odyssey of Voyager 2, by Joel Davis. Atheneum, 1987.

Rings: Discoveries from Galileo to Voyager, by James Elliot and Richard Kerr. The MIT Press, 1984.

Planets X and Pluto, by William Graves Hoyt. University of Arizona Press, 1980.

Planets Beyond: Discovering the Outer Solar System (revised printing), by Mark Littman. John Wiley and Sons, 1990.

Uranus: The Planet, Rings, and Satellites, by Ellis D. Miner. Ellis Horwood Publishers, 1990.

Distant Encounters: The Exploration of Jupiter and Saturn, by Mark Washburn. Harcourt Brace Jovanovich, Publishers, 1983.

COMETS, ASTEROIDS, METEORITES

Cosmic Debris: Meteorites in History, by John G. Burke. University of California Press, 1986.

Rendezvous in Space: The Science of Comets, by John C. Brandt and Robert D. Chapman. W.H. Freeman and Company, 1992.

Thunderstones and Shooting Stars: The Meaning of Meteorites, by Robert T. Dodd. Harvard University Press, 1986.

The Search for Our Beginning, by Robert Hutchison. Oxford University Press, 1983.

Asteroids: Their Nature and Utilization, by Charles T. Kowal. John Wiley and Sons, 1988.

Meteorites and Their Parent Bodies, by Harry Y. McSween, Jr. Cambridge University Press, 1987.

Comet, by Carl Sagan and Ann Druyan. Simon and Schuster, 1985.

The Mystery of Comets, by Fred L. Whipple. Smithsonian Institution Press, 1985.

Comets: A Chronological History of Observation, Science, Myth, and Folklore, by Donald K. Yeomans. John Wiley and Sons, 1991.

SOURCES OF ILLUSTRATIONS

Illustrations drawn by Ian Worpole and April Pahl

Frontispiece
Oronce Finé, *Le spher du monde*, Paris, 1549. The Houghton Library, Harvard University

Opposite page 1
Macduff Everton

Page 1
Roger Ressmeyer/Starlight Photo Agency

Page 6
Giraudon/Art Resource

Page 7
Nicolaus Copernicus, De revolutionibus orbium coelestium, 1543. Library of Congress

Page 8
Erich Lessing/Art Resource

Page 9
Erich Lessing/Art Resource

Page 15
National Portrait Gallery, London

Page 18
NASA/Jet Propulsion Laboratory

Page 19
NASA/Jet Propulsion Laboratory

Page 21
top right, Leon Golub, IBM Research and Smithsonian Astrophysical Observatory
bottom, NASA/Jet Propulsion Laboratory

Page 27
NASA/Ames Research Center

Page 30
NOAO

Page 35
NASA/Jet Propulsion Laboratory

Page 39
NASA/Jet Propulsion Laboratory

Page 40
NASA/Jet Propulsion Laboratory

Page 42
NASA/Jet Propulsion Laboratory

Page 44
NASA/Jet Propulsion Laboratory

Page 45
NASA/Jet Propulsion Laboratory

Page 46
NASA/Jet Propulsion Laboratory

Page 50
NASA

Page 51
NASA/Johnson Space Center

Page 53
Scala/Art Resource

Page 54
Library of Congress

Page 55
NASA/Johnson Space Center

Page 56
NASA/Johnson Space Center

Page 57
USGS, Flagstaff, Arizona

Page 58
Lick Observatory

Page 59
NASA/Johnson Space Center

Page 60
NASA/Johnson Space Center

Page 61
NASA/Johnson Space Center

Page 63
NASA/Johnson Space Center

Page 69
NASA

Page 71
MoMA Film Stills

Page 72
NASA/Jet Propulsion Laboratory

Page 73
NASA

Page 74
NASA/Jet Propulsion Laboratory

Page 75
NASA/Jet Propulsion Laboratory

Page 78
William Benz

Page 80
NASA/Jet Propulsion Laboratory

Page 81
NASA

Page 84
top, NASA

Pages 84–85
NASA

Page 87
NASA/Jet Propulsion Laboratory

Page 90
NASA/Jet Propulsion Laboratory

Page 92
NASA/Jet Propulsion Laboratory

Page 93
Dave Pieri/Jet Propulsion Laboratory

Page 94
NASA/Jet Propulsion Laboratory

Page 95
NASA/Jet Propulsion Laboratory

Page 96
NASA/Jet Propulsion Laboratory

Page 100
Art Wolfe

Page 102
NASA/Jet Propulsion Laboratory

Page 105
NASA/Johnson Space Center

Page 109
E. A. Shinn, USGS

Page 110
Heather Angel/Biofotos

Page 113
Jim Lukoski/Black Star

Page 114
USGS, Flagstaff, Arizona

Page 115
NASA

Page 117
Lowell Observatory

Page 118
Lowell Observatory

Page 121
Mary Dale-Bannister, Washington University

Page 122
Mary Dale-Bannister, Washington University

Page 124
NASA/Jet Propulsion Laboratory

Page 126
NASA/Jet Propulsion Laboratory

Page 127
NASA/Jet Propulsion Laboratory

Page 129
NASA/Jet Propulsion Laboratory

Page 131
NASA/Jet Propulsion Laboratory

Page 133
USGS, Flagstaff, Arizona

Page 138
USGS, Flagstaff, Arizona

Page 139
USGS, Flagstaff, Arizona

Page 141
NASA/Jet Propulsion Laboratory

Page 142
NASA/Jet Propulsion Laboratory

Page 145
Michael Carroll

Page 148
Mary Dale-Bannister, Washington University

Page 149
Roger Bourke/Planetary Report

Page 150
NASA/Jet Propulsion Laboratory

Page 151
NASA/Johnson Space Center

Page 153
The Pierpont Morgan Library, New York. MA 1064

Page 159
NASA/Jet Propulsion Laboratory

Page 160
NASA/Jet Propulsion Laboratory

Page 161
USGS, Flagstaff, Arizona

Page 163
NASA/Jet Propulsion Laboratory

Page 164
NASA/Jet Propulsion Laboratory

Page 165
NASA/Jet Propulsion Laboratory

Page 166
NASA/Jet Propulsion Laboratory

Page 167
NASA/Jet Propulsion Laboratory

Page 168
NASA/Jet Propulsion Laboratory

Page 171
NASA/Jet Propulsion Laboratory

Page 173
NASA/Jet Propulsion Laboratory

Page 174
NASA/Jet Propulsion Laboratory

Page 175
NASA/Jet Propulsion Laboratory

Page 177
NASA/Jet Propulsion Laboratory

Page 180
ROE/AAT Board, 1986

Page 181
European Space Agency

Page 183
Field Museum of Natural History

Page 184
Field Museum of Natural History

Page 185
NASA/Johnson Space Center

Page 187
Chip Clark

Page 192
Steven Ostro/Jet Propulsion Laboratory

Page 195
NASA/Jet Propulsion Laboratory

Page 196
NASA/Jet Propulsion Laboratory

Page 197
Planetary Arts and Sciences

Page 198
Diebold-Schilling-Chronik, 1513. Zentralbiblioteque Luzern/Korporation Luzern

Page 201
Uwe Fink, University of Arizona

Page 202
Max Planck Institut für Aeronomie

Page 205
Donald Brownlee, University of Washington

Page 209
NOAO

Page 210
"Red Star in the Clouds" by Dürer (back of panel "St. Jerome in the Wilderness.") By permission of the Syndics of the Fitzwilliam Museum, Cambridge

Page 214
Royal Tyrrell Museum/Alberta Culture and Multiculturalism

Page 216
Bruce F. Bohor/USGS

Page 218
Sovfoto

Page 219
Meteor Crater Enterprises, Arizona

Page 220
Radarsat International

Page 221
Field Museum of Natural History

Page 223
NASA/Jet Propulsion Laboratory

INDEX

Other books in the Scientific American Library Series

POWERS OF TEN
by Philip and Phylis Morrison and the Office of Charles and Ray Eames

HUMAN DIVERSITY
by Richard Lewontin

THE DISCOVERY OF SUBATOMIC PARTICLES
by Steven Weinberg

FOSSILS AND THE HISTORY OF LIFE
by George Gaylord Simpson

ON SIZE AND LIFE
by Thomas A. McMahon and John Tyler Bonner

FIRE
by John W. Lyons

SUN AND EARTH
by Herbert Friedman

ISLANDS
by H. William Menard

DRUGS AND THE BRAIN
by Solomon H. Snyder

THE TIMING OF BIOLOGICAL CLOCKS
by Arthur T. Winfree

EXTINCTION
by Steven M. Stanley

EYE, BRAIN, AND VISION
by David H. Hubel

THE SCIENCE OF STRUCTURES AND MATERIALS
by J. E. Gordon

SAND
by Raymond Siever

THE HONEY BEE
by James L. Gould and Carol Grant Gould

ANIMAL NAVIGATION
by Talbot H. Waterman

SLEEP
by J. Allan Hobson

FROM QUARKS TO THE COSMOS
by Leon M. Lederman and David N. Schramm

SEXUAL SELECTION
by James L. Gould and Carl Grant Gould

THE NEW ARCHAEOLOGY AND THE ANCIENT MAYA
by Jeremy A. Sabloff

A JOURNEY INTO GRAVITY AND SPACETIME
by John Archibald Wheeler

SIGNALS
by John R. Pierce and A. Michael Noll

BEYOND THE THIRD DIMENSION
by Thomas F. Banchoff

DISCOVERING ENZYMES
by David Dressler and Huntington Potter

THE SCIENCE OF WORDS
by George A. Miller

ATOMS, ELECTRONS, AND CHANGE
by P. W. Atkins

VIRUSES
by Arnold J. Levine

DIVERSITY AND THE TROPICAL RAIN FOREST
by John Terborgh

(continued)

STARS
by James B. Kaler

EXPLORING BIOMECHANICS
by R. McNeill Alexander

CHEMICAL COMMUNICATION
by William C. Agosta

GENES AND THE BIOLOGY OF CANCER
by Harold E. Varmus and Robert A. Weinberg

SUPERCOMPUTING AND THE TRANSFORMATION OF SCIENCE
by William J. Kaufmann III and Larry L. Smarr

MOLECULES AND MENTAL ILLNESS
by Samuel H. Barondes